Insect Evolution in an Amberiferous and Stone Alphabet

Insect Evolution in an Amberiferous and Stone Alphabet

Proceedings of the 6th International Congress on Fossil Insects, Arthropods and Amber

Edited by

Dany Azar, Michael S. Engel, Edmund Jarzembowski,
Lars Krogmann, André Nel and Jorge Santiago-Blay

TUTA SUB AEGIDE PALLAS · 1683 ·

BRILL

LEIDEN · BOSTON
2013

Library of Congress Cataloging-in-Publication Data

International Congress on Fossil Insects, Arthropods and Amber (6th : 2013 : Jubayl, Lebanon)
Insect evolution in an amberiferous and stone alphabet : proceedings of the 6th International Congress on Fossil Insects, Arthropods and Amber / edited by DanyAzar, Michael S. Engel, Edmund Jarzembowski, Lars Krogmann, AndreNel and Jorge Santiago-Blay.
pages cm
Includes bibliographical references and index.
ISBN 978-90-04-21070-7 (hardback : acid-free paper) -- ISBN 978-90-04-21071-4 (e-book) 1. Insects, Fossil--Congresses. 2. Amber fossils--Congresses. 3. Geology, Stratigraphic--Cretaceous--Congresses. 4. Geology, Stratigraphic--Jurassic--Congresses. 5. Insects--Evolution--Congresses. I. Azar, Dany, 1973- editor of compilation. II. Title.

QE831.I58 2013
565'.7--dc23

2013022250

This publication has been typeset in the multilingual "Brill" typeface. With over 5,100 characters covering Latin, IPA, Greek, and Cyrillic, this typeface is especially suitable for use in the humanities. For more information, please see www.brill.com/brill-typeface.

ISBN (hardback): 978-90-04-21070-7
ISBN (e-book): 978-90-04-21071-4

This book is printed on acid-free paper.

Contents

Preface

Humanity seeks to explain itself and the world in which it lives. We observe, seek patterns, and classify things around us, striving to contextualize and understand our existence, and ultimately to find meaning therein. An integral part of this quest is to comprehend what brought us to our present lives. For instance, the translation and study of ancient languages and literatures informs us of the human condition, illuminating the daily toils of our distant predecessors and helping us to better interpret their perceptions of the world in which they lived and fundamentally explain how we today choose to perceive our own realities. Through words carved in rock or lovingly gathered in scrolls or codices, sublime etchings or sculptures of stone or wood, or transcendent renditions painted with and on a myriad of materials, we decipher the life of those who came before us. We work tirelessly to resurrect their lives, to contextualize their existence, and to reveal a society and culture simultaneously alien to the one in which we live and yet historically and genealogically consanguineous.

In the same manner, palaeoentomology seeks to read the record of the lives of insects and their allies. As with the study of our own condition, if we desire to truly fathom the successes and importance of today's arthropods, then we must have a deep appreciation and understanding of their distant past. Where historians and archaeologists have their cuneiform tablets or marble sculptures, the palaeoentomologist finds the lives of his subject 'written' in compressions, impressions, and inclusions, a metaphorical alphabet of amber and stone. This historical record may also be as aesthetically stunning as anything Lysippos could have created. Who among us has not gazed in awe at the beauty of metallic beetles recovered from Grube Messel, Germany and the last global greenhouse, at delicate and life-like lacewings entombed in amber from the 'Age of the Dinosaurs', or stared in wonder at the weighty wings of massive griffenflies cast in stone and challenging us to imagine them aloft in the distant skies of the Carboniferous and Permian? These remains not only stir our scientific curiosities but animate our passions. These are the various descendants of life's greatest diversification, of a lineage that traces its ancestry more than 400 million years, witnessing cataclysms of global intensity as well as the world's blossoming. Throughout it all, insects have existed, persisted, flourished and perished, and, in many instances, were intimately involved in the similar experiences of countless more lineages, including our own. Our evolution, physical and cultural, is inextricably connected with that of insects whether as pests or benefactors. It is only appropriate that some of us devote our lives to more fully understanding theirs.

Michael S. Engel
Lawrence, KS, USA
February 2013

Introduction

Insect Evolution in an Amberiferous and Stone Alphabet

Dany Azar[a], Michael S. Engel[b], Edmund Jarzembowski[c], Lars Krogmann[d], André Nel[e] and Jorge Santiago-Blay[f]

[a]Lebanese University, Faculty of Sciences II, Department of Natural Sciences P.O. Box 26110217, Fanar–Matn, Lebanon; e-mail: azar@mnhn.fr
[b]Division of Entomology, Natural History Museum, and Department of Ecology & Evolutionary Biology, 1501 Crestline Drive, Suite 140, University of Kansas, Lawrence KS 66045, USA; e-mail: msengel@ku.edu
[c]Key State Laboratory of Palaeobiology and Stratigraphy, Nanjing Institute of Geology and Palaeontology, Chinese Academy of Sciences, 39 East Beijing Road, Nanjing 210008 P.R. China; e-mail: jarzembowski2@live.co.uk
[d]State Museum of Natural History Stuttgart, Entomology, Rosenstein 1, D-70191 Stuttgart Germany; e-mail: lars.krogmann@smns-bw.de
[e]CNRS UMR 7205, Muséum National d'Histoire Naturelle, CP 50, Entomologie 45 Rue Buffon, F-75231 Paris, France; e-mail: anel@mnhn.fr
[f]Department of Paleobiology, National Museum of Natural History, Smithsonian Institution, Washington, DC 20560, USA; e-mail: blayjorge@gmail.com

---

The 6th International Congress on Fossil Insects, Arthropods and Amber

During the past 20 years, palaeoentomology has undergone remarkable progress in parallel to the increasing number of amber outcrops due to the international growth of scientific awareness and interest in this subject. There is no doubt that "Jurassic Park" in 1993, the famous American science fiction adventure thriller film directed by Steven Spielberg, and based on the eponymous novel of Michael Crichton, played a noticeable role in making amber more popular. Before this date interest in amber was mainly restricted to the Caribbean region (the Dominican Republic), Chiapas, Mexico and, especially, the Baltic countries (with a great tradition of amber appreciation), although it was recognized that amber occurred in various localities world-wide.

The Six Congresses

The first of the 6 congresses on Fossil Insects, Arthropods and Amber was the First Palaeoentomological Conference in Russia (Moscow, 1998)—not surprising in view of the presence of the Arthropod Laboratory there. It was also the same year as the First World Congress on Amber Inclusions in Vitoria-Gasteiz (Basque Country, Spain)—also not surprising in view of the discovery of Cretaceous amber there, although younger than Lebanese amber. Before that we had the Fossil Insect Network (1996) based in Strasbourg, France with its exceptionally preserved Triassic fossils, and the headquarters of the European Science Foundation; but that was limited to the European Union, although producing the free on-line *Meganeura* newsletter. Before that there were only some national meetings, notably in Germany and, later, in Poland, the home of another newsletter, *Wrostek – Inclusion*. The new international conference spirit meant that we were to meet at least every four years henceforth and produce a conference volume: it turned out to be nearer three years due to its popularity.

Gondwana responded to this Laurasian initiative in 2000 when the First International Meeting on Palaeoarthropodology was held in Ribeirao Preto (near Sao Paulo, Brazil) targeting non-marine arthropods and also embracing the first Brazilian and South American palaeoarthropod symposia. Sadly, the palaeontology department there and its head, Rafael Martins-Neto, are no longer with us.

The second conference, now dubbed a congress, was in Poland (Kraków) in 2001 where the International Palaeoentomological Society (IPS) was born amidst a Baltic amber exhibition under construction.

By the third congress in South Africa (Pretoria) common interests resulted in a single, combined congress on insects/arthropods/amber under the auspices of the IPS and Fossils x3 (for short) was born.

We returned to Spain in 2007 with the fourth congress which launched the new hardcopy journal *Alavesia* but which was, however, subsequently hit by the European Credit Crunch from which we are still recovering.

In 2010 we had our first congress in Asia at Beijing in China, not too far from Cretaceous volcanic lakes with exceptionally preserved compression fossils.

And now, Byblos (home of the first alphabet) in Lebanon, the first Middle Eastern congress in a country with a great Cretaceous amber heritage and a couple of challenges- will the registration of the Society be finally achieved this year (in France) and a new economic supporter for *Alavesia* found?

This congress was held in the oldest continuously-inhabited city in the world, the biblical city of Byblos, Lebanon, 14–18 April 2013, under the High Patronage of His Excellency, the President of the Republic of Lebanon, General Michel Sleiman, and gathered about 110 participants from 19 different countries (Argentina, Australia, Belgium, China, Czech Republic, Denmark, France, Germany, Mexico, Morocco, Poland, Russia, Slovakia, Spain, Syria, Ukraine, United Kingdom, United States of America and Lebanon). 32 original papers covering a wide spectrum of topics relating to insect systematics and phylogeny, amber and its preparatory techniques, geology and palaeoenvironment were accepted as proceedings of this congress, among which 11 have been published in *Insect Systematics & Evolution* (Cai & Huang 2013a,b; Khramov 2013; Żyła & Wegierek 2013; Kania et al. 2013; Kirejtshuk & Nel 2013; Nel et al. 2013; Prokop et al. 2013; Fang et al. 2013a,b; Ross & Engel 2013), nine in volume 6(1–2) of *Terrestrial Arthropod Reviews* (Jarzembowski et al. 2013; Nohra et al. 2013; Veltz et al. 2013; Petrulevicius 2013; Baranov & Perkovsky 2013; Perkovsky & Rasnitsyn 2013; Kirejtshuk & Azar 2013; Moseyko & Kirejtshuk 2013; Ain Malak et al. 2013) and 12 in this book (Nel & Azar 2013; Azar et al. 2013; Choufani et al. 2013; Drohojowska & Szwedo 2013; Herczek et al. 2013; Szwedo et al. 2013; Coty et al. 2013; Vršanský & Makhoul 2013; Krzemiński et al. 2013; Azar et al. 2013; Sidorchuk 2013; Kirejtshuk & Poinar 2013). About 45 new fossil taxa resulted from this congress. After the closure ceremony of the congress, His Excellency, the President of the Republic of Lebanon, General Michel Sleiman welcomed participants in the Presidential Palace (Fig. 1), and in his address he promised to do his best to initiate the creation of a Lebanese National Museum of Natural History.

During the congress, the archeological sites of Byblos and the wonderful caves of Jeita were visited. The field trip included a visit to Lower Cretaceous amber sites at Roum-Aazour-Homsiyyeh, Jouar Ess-Souss (Bkassine) and Hammana (in Southern and Central Lebanon); the Jurassic amber site of Aintourine and the natural reserve of Ehden Forest (Northern Lebanon).

State of the art and perspectives in the research on fossil insects and amber: a plea for integration

Due to the spectacular recent progress in molecular phylogeny and of the beginning of the "age of phylogenomics", researchers may be tempted to think that the use of morphological characters to generate phylogenies could be outdated. The situation could also be considered worse

concerning the use of fossils because of the fragmentary state of the available material. The actual situation is exactly the opposite. Morphological phylogenies are essential to test and enrich the results obtained by the new molecular tools because they provide independent tests of phylogenetic hypotheses. In fact, it is a mistake to consider *a priori* that a particular set of characters will lead to the 'Holy Grail' of phylogeny, i.e. to the exact and complete reconstruction of the history of life. We can attempt to approach this result asymptotically only through the integration of the available data into a process of constant and respectful criticism. The phylogenetic classification of fossil taxa is an essential part of the reconstruction of the unique evolutionary transformations that have affected the different organisms. It is, therefore, logical to use all available data (molecular, recent and fossil morphology) and to compare the resulting phylogenies- just like human historians use recent, written and archaeological data to debate the human past.

The participants in the 6th International Congress on Fossil Insects, Arthropods and Amber with His Excellency the President of the Republic of Lebanon, General Michel Sleiman. From left to right: 1, Mr. Paul Nascimbene (USA); 2, Mrs. Denise Laneve (USA); 3, Dr. Isabelle Veltz (France); 4, Dr. Jean,Claude Paicheler (France); 5, Dr. Julián Petrulevičius (Argentina); 6, Mrs. Pierrette Gex (France); 7, Dr. Bo Wang (China); 8, Dr. Jacek Szwedo (Poland); 9, Dr. Sarah Martin (Australia); 10, Prof. Piotr Węgierek (Poland); 11, Prof. Aleksander Herczek (Poland); 12, Dr. Yan Fang (China); 13, Dr. Iwona Kania (Poland); 14, Prof. Wiesław Krzemiński (Poland); 15, Prof. Alexandr Rasnitsyn (Russia); 16, Prof. Ewa Krzemińska (Poland); 17, Dr. Ekaterina Sidorchuk (Russia); 18, Mr. Dimitry Vorontsov (Russia); 19, Dr. Youmna Salamé (Lebanon); 20, Dr. Enrique Peñalver (Spain); 21, Dr. Alexander Khramov (Russia); 22, Mrs. Joanna Choufani (Lebanon); 23, Prof. Alexander Kirejtshuk (Russia); 24, Prof. Alexey Solodovnikov (Denmark); 25, Mr. Georges Kachacha (Lebanon); 26, Dr. Olivier Béthoux (France); 27, Dr. Andrew Ross (UK); 28, Prof. Daniel Bickel (Australia); 29, Prof. Ali Mneymneh (Lebanon); 30, His Excellency the President of the Republic of Lebanon, General Michel Sleiman (Lebanon); 31, Dr. Georges Rahbani (Lebanon); 32, Dr. Daran Zheng (China); 33, Prof. Dany Azar (Lebanon); 34, Mr. Chenyang Cai (China); 35, Mr. Youssef Nohra (Lebanon); 36, Prof. Diying Huang (China); 37, Mr. Kamil Ziadé (Lebanon); 38, Prof. Antoine El-Samrani (Lebanon); 39, Prof. Ed Jarzembowski (UK); 40, Prof. Evgeny Perkovsky (Ukraine); 41, Dr. Vladimir Blagoderov (UK); 42, Dr. Ricardo Pérez. De La Fuente (Spain); 43, Dr. Evgeny Yan (China); 44, Mrs. Biddy Jarzembowski (UK); 45, Dr. Christel Hoffeins (Germany); 46, Mr. Heins Hoffeins (Germany); 47, Mrs. Kornelia Skibińska (Poland); 48, Prof. Yuri Popov (Russia); 49, Dr. Mónica Morayma Solórzano Kraemer (Mexico); 50, Mrs. Alba Sánchez García (Spain); 51, Prof. Dong Ren (China); 52, Prof. Michael S. Engel (USA); 53, Mrs. Martina Pecharová (Czech Republic); 54, Prof. Alexander Schmidt (Germany); 55, Prof. Jakub Prokop (Czech Republic).

In the case of the Hexapoda, fossils are generally not so fragmentary and incomplete as neontologists frequently claim. The complete habitus of these organisms is frequently quite well preserved. But while researchers working on vertebrate phylogeny integrate regularly the fossil taxa in their phylogenies and classifications, many neoentomologists still consider fossil insects as uninformative and useless—like historians decrying archaeology. Fortunately, the situation is progressively changing thanks to the impressive amount of new data that became available after the end of the Cold War during the period 1990–2013, viz., descriptions of new fossil taxa based on exquisitely preserved material, successful attempts at integrating fossil insects in morphological phylogenies, and more frequent use of fossils to calibrate the molecular dating of clades. Until the last five years, the study of fossil insects in amber and rocks was only done by traditional optical and stereomicroscopy. Today, with the dramatic advances in technology, a great step towards the future has been made with the use of new tools of exploration and imaging, even in opaque material. These tools include X-ray synchrotron tomography or the new generation of micro-CT scanners that are beginning to become more precise and available, although still expensive. The new technology does not allow only the study of external morphology of the inclusions, but also their tiny internal anatomy (genitalia, mouthparts, etc.), and all this in a safe, non-destructive way. Consequently, many fossils can now provide a quantity of morphological information comparable to that obtained from their modern relatives. Furthermore, many fossils can be considered as keystone taxa for phylogenetic analyses because they show characters, or combinations of characters, and distributions no longer present in the living world. Lastly, the recent studies have shown concretely what was suspected a long time ago: the recent biodiversity is only a fraction of the billions of species that have lived on Earth in the deep past. The researcher who neglects this information is simply unable to reconstruct the history of life. The father of cladistic phylogeny, Willi Hennig, who would have turned 100 on 20 April 2013, just two days after this congress, was aware of this situation when he studied, described and tried to integrate fossil flies in his phylogenetic analyses so that phylogeny was not a mere statement of current relationships, but a statement about ancestor-descendant relationships based on a consideration of the inherited past.

Outlook: Our collective perspective

Due to the growing interest in palaeoentomology and globalization, several serious multidisciplinary and collaborative scientific teams have been formed in many countries, and many new professionals have completed doctorates or are in the process of completing them. What is also really marvellous is the fact that with the present technologies of communication, most of these scientists from all over the world are collaborating remotely in international teams. Consequently, a relatively large number of high quality publications are produced annually. Moreover, fossil material is more and more attractive to neoentomologists, gradually integrating with palaeontological teams, augmenting the quality of research and our knowledge of the historical evolution of the groups studied.

We sincerely believe that scientific research on fossil insects and amber will continue to flourish and prosper in the future. We are quite sure that surprising results will emerge globally in tandem with the development of technology and the growing interest in this subject. In a short time, the modern renaissance of palaeoentomology has already shown that people from different cultures and nationalities can successfully co-operate in this scientific quest, in contrast to the colonial world of the founding fathers of the science in past centuries.

Acknowledgements

We thank all the members of the Organizing Committee of the 6th International Congress on Fossil Insects, Arthropods and Amber. We thank as well all the persons and institutes (especially the Lebanese University, the Municipalities of Byblos and Bkassine, and the Ministry of Tourism in Lebanon) that made possible the venue of the congress. We thank His Excellency General Michel Sleiman, President of the Republic of Lebanon, for his High Patronage of the congress; we thank him as well for his warm welcome of the delegates in the Presidential Palace.

References

Ain Malak, M., Salamé, Y. & Azar, D. (2013) New phlebotomine flies from Burmese amber (Diptera: Psychodidae: Phlebotominae). *Terrestrial Arthropods Reviews* 6: 81–101.

Azar, D. & Nel, A. (2013) A new beaded lacewing from a new Lower Cretaceous amber outcrop in Lebanon (Neuroptera: Berothidae). In: Azar, D., Engel, M.S., Jarzembowski, E., Krogmann, L., Nel, A. & Santiago-Blay, J. (Eds) *Insect Evolution in an Amberiferous and Stone Alphabet*. Brill, Leiden, pp. 111–129.

Azar, D., Nel, A. & Perkovsky, E. (2013) A new *Sycorax* species from Eocene Ukrainian Rovno amber (Diptera: Psychodida: Sycoracinae). In: Azar, D., Engel, M.S., Jarzembowski, E.,

Krogmann, L., Nel, A. & Santiago-Blay, J. (Eds) *Insect Evolution in an Amberiferous and Stone Alphabet*. Brill, Leiden, pp. 27–46.

Azar, D., Nohra, Y., Iskandar, D. & Gèze, R. (2013) Fanar, a "dream" Lebanese Lower Cretaceous amber outcrop, dissipated... In: Azar, D., Engel, M.S., Jarzembowski, E., Krogmann, L., Nel, A. & Santiago-Blay, J. (Eds) *Insect Evolution in an Amberiferous and Stone Alphabet*. Brill, Leiden, pp. 175–186.

Baranov, V. & Perkovsky, E. (2013) Chironomids (Diptera: Chironomidae) in the Sakhalinian amber (Eocene). Part I. *Terrestrial Arthropods Reviews* **6**: 61–69.

Cai, C & Huang, D. (2013a) Discussion on the systematic position of the oxyteline rove beetle *Anotylus archaicus* Yue, Makranczy & Ren, 2012 (Coleoptera: Staphylinidae). *Insect Systematics & Evolution* **44**: 203–212.

Cai, C. & Huang, D. (2013b) *Mesocaprophilus clavatus*, a New oxyteline rove beetle (Coleoptera: Staphylinidae) from the Early Cretaceous of China. *Insect Systematics & Evolution* **44**: 213–220.

Choufani, J., Perrichot, V., Girard, V., Garrouste, R., Azar, D. & Nel, A. (2013) Two new biting midges of modern type from the Santonian amber of France (Diptera: Ceratopogonidae). In: Azar, D., Engel, M.S., Jarzembowski, E., Krogmann, L., Nel, A. & Santiago-Blay, J. (Eds) *Insect Evolution in an Amberiferous and Stone Alphabet*. Brill, Leiden, pp. 73–95.

Coty, D., Salamé, Y., Nel, A. & Azar, D. (2013) A new psychodid fly from Mexican amber (Diptera; Psychodidae). In: Azar, D., Engel, M.S., Jarzembowski, E., Krogmann, L., Nel, A. & Santiago-Blay, J. (Eds) *Insect Evolution in an Amberiferous and Stone Alphabet*. Brill, Leiden, pp. 13–25.

Drohojowska, J. & Szwedo, J. (2013) *Gapenus rhinariatus* gen. sp. n. from the Lower Cretaceous amber of Lebanon. In: Azar, D., Engel, M.S., Jarzembowski, E., Krogmann, L., Nel, A. & Santiago-Blay, J. (Eds) *Insect Evolution in an Amberiferous and Stone Alphabet*. Brill, Leiden, pp. 99–110.

Fang, Y. Zhang, H., Wang, B. & Zheng, D. (2013a) A new cockroach (Blattodea: Caloblattinidae) from the Upper Triassic Xujiahe Formation of Sichuan Province, southwestern China. *Insect Systematics & Evolution* **44**: 167–174.

Fang, Y., Zhang, H., Wang, B. & Zheng, D. (2013b) A new Chifengiine species (Orthoptera: Prophalangopsidae) from the Lower Cretaceous Yixian Formation of Liaoning, China. *Insect Systematics & Evolution* **44**: 141–147.

Herczek, A., Popov, Y. & Perkovski, E. (2013) Another new representative of the isometopine genus *Archemyiomma* (Hemiptera: Heteroptera: Miridae) from Late Eocene Rovno (Ukraine) amber. In: Azar, D., Engel, M.S., Jarzembowski, E., Krogmann, L., Nel, A. & Santiago-Blay, J. (Eds) *Insect Evolution in an Amberiferous and Stone Alphabet*. Brill, Leiden, pp. 47–54.

Jarzembowski, E.A., Yan, E.V., Wang, B. & Zhang, H. (2013) Ommatine beetles (Insecta: Coleoptera) from the Lower Cretaceous of northeast China and southern England. *Terrestrial Arthropods Reviews* **6**: 135–161.

Kania, I., Krzemiński, W. & Azar, D. (2013) The oldest representative of *Helius* Lepeletier & Serville 1828 (Limoniidae, Diptera) from Lebanese amber (Early Cretaceous). *Insect Systematics & Evolution* **44**: 231–238.

Khramov, A. (2013) New mantidflies (Neuroptera: Mantispidae) from the Upper Jurassic of Kazakhstan. *Insect Systematics & Evolution* **44**: 221–230.

Kirejtshuk, A.G. & Azar, D. (2013) Current knowledge of Coleoptera (Insecta) from the Lower Cretaceous Lebanese amber. *Terrestrial Arthropods Reviews* **6**: 103–134.

Kirejtshuk, A.G. & Nel, A. (2013) Current knowledge of Coleoptera (Insecta) from the Lowermost Eocene Oise amber. *Insect Systematics & Evolution* **44**: 175–202.

Kirejtshuk, A.G. & Poinar, G. (2013) On the systematic position of the genera *Lepiceroides* gen.n. and *Haplochelus* with notes on the taxonomy and phylogeny of the Myxophaga (Coleoptera). In: Azar, D., Engel, M.S., Jarzembowski, E., Krogmann, L., Nel, A. & Santiago-Blay, J. (Eds) *Insect Evolution in an Amberiferous and Stone Alphabet.* Brill, Leiden, pp. 55–69.

Krzemiński, W., Azar, D. & Skibińska, K. (2013) *Nannotanyderus ansorgei* sp. n. - the first member of the family Tanyderidae from the Lebanese amber (Lower Cretaceous). In: Azar, D., Engel, M.S., Jarzembowski, E., Krogmann, L., Nel, A. & Santiago-Blay, J. (Eds) *Insect Evolution in an Amberiferous and Stone Alphabet.* Brill, Leiden, pp. 131–143.

Moseyko, A.G. & Kirejtshuk, A.G. (2013) On the taxonomic position of new taxa of the subfamily Eumolpinae (Coleoptera, Chrysomelidae) from Baltic amber. *Terrestrial Arthropods Reviews* **6**: 163–172.

Nel, A., Krzeminski, W. & Szwedo, J. (2013) *Elektroeuphaea* gen.n. - the oldest representative of the modern Epallaginae (Odonata: Zygoptera: Epallagidae). *Insect Systematics & Evolution* **44**: 129–140.

Nohra, Y., Azar, D., Gèze, R., Maksoud, S., El-Samrani, A. & Perrichot, V. (2013) New Jurassic amber outcrops from Lebanon. *Terrestrial Arthropods Reviews* **6**: 27–51.

Perkovsky, E. & Rasnitsyn, A. (2013) Biting midges (Diptera, Ceratopogonidae) in amber forest communities based on analysis of syninclusions in the Late Eocene Rovno amber. *Terrestrial Arthropods Reviews* **6**: 71–80.

Petrulevičius, J.F. (2013) Palaeoenvironmental and palaeoecological implications from body fossils and ovipositions of Odonata from the Eocene of Patagonia, Argentina. *Terrestrial Arthropods Reviews* **6**: 53–60.

Prokop, J., Pecharová, M., Roques, P. & Nel, A. (2013) A new genus and species of Breyeriidae from the Upper Carboniferous of Nord-Pas-de-Calais, France supplemented by wings of immature stages (Insecta: Palaeodictyoptera). *Insect Systematics & Evolution* **44**: 117–128.

Ross, A.J. & Engel, M.S. (2013) The first diplatyid earwig in Tertiary amber (Dermaptera: Diplatyidae): A new species from Miocene Mexican amber. *Insect Systematics & Evolution* **44**: 157–166.

Sidorchuk, E.S. (2013) New technique for preparation of small-sized amber samples with application to mites. In: Azar, D., Engel, M.S., Jarzembowski, E., Krogmann, L., Nel, A. & Santiago-Blay, J. (Eds) *Insect Evolution in an Amberiferous and Stone Alphabet.* Brill, Leiden, pp. 189–201.

Szwedo, J., Azar, D. & Nohra, Y. (2013) First record of Perforissidae from the Lower Cretaceous Lebanese amber (Hemiptera: Fulgoromorpha: Fulgoroidea). In: Azar, D., Engel, M.S., Jarzembowski, E., Krogmann, L., Nel, A. & Santiago-Blay, J. (Eds) *Insect Evolution in an Amberiferous and Stone Alphabet.* Brill, Leiden, pp. 145–163.

Veltz, I., Paicheler, J.-C., Maksoud, S., Gèze, R. & Azar, A. (2013) Context and genesis of the Lebanese ambriferous palaeoenvironments at the Jurassic-Cretaceous transition. *Terrestrial Arthropods Reviews* **6**: 11–26.

Vršanský, P. & Makhoul, E. (2013) *Mieroblattina pacis* gen. et sp. nov. - Upper Cretaceous cockroach (Blattida: Mesoblattinidae) from Nammoura limestone of Lebanon. In: Azar, D., Engel, M.S., Jarzembowski, E., Krogmann, L., Nel, A. & Santiago-Blay, J. (Eds) *Insect Evolution in an Amberiferous and Stone Alphabet.* Brill, Leiden, pp. 167–172.

Żyła, D. & Węgirek, P. (2013) The oldest Palaeoaphididae (Insecta: Hemiptera: Aphidomorpha) from Late Jurassic/Early Cretaceous locality of Khotont (Mongolia). *Insect Systematics & Evolution* 44: 149–155.

PART ONE

INSECTS FROM CAENOZOIC AMBER

A new psychodid fly from Mexican amber (Diptera; Psychodidae)

Davide Coty[a], Youmna Salamé[b], André Nel[a] and Dany Azar[b,*]

[a]CNRS UMR 7205, Muséum National d'Histoire Naturelle, CP 50, Entomologie, 45 rue Buffon, F-75005 Paris, France
[b]Lebanese University, Faculty of Sciences II, Department of Natural Sciences, Fanar, P.O. Box 26110217, Fanar–Matn, Lebanon
*Corresponding author, e-mail: azar@mnhn.fr

Abstract

Succinarisemus totolapensis sp. n., a new psychodine fly from Mexican amber, is characterised, described and illustrated. It represents the second representative and fossil belonging to *Succinarisemus*. This discovery enriches our knowledge of the palaeodiversity of this peculiar fossil psychodine genus know elderly only from Dominican amber. Palaeobiogeographical data are discussed.

Keywords

Psychodinae; Mexican amber; Oligocene/Miocene; sp. n.; *Succinarisemus mexicanus*

Introduction

The family Psychodidae (Diptera: Nematocera) are small flies with short, hairy bodies and wings that give them a moth-like appearance. The adults have long antennae, containing 12–16 segments, and each segment is bulbous with a ring of long hairs (Quate & Vockeroth 1981; Wagner 1997). Their wings are leaf-shaped, either slender or broad, with 9–10 longitudinal veins and no cross-veins in the distal two-thirds of each wing (Quate & Vockeroth 1981; Wagner 1997). Adult Psychodidae are commonly nocturnal, associated with damp habitats, and more common and diverse in intertropical regions (Grimaldi & Engel 2005). Some Psychodidae (such as members of Phlebotominae subfamily) are blood feeders, and important vectors of several infectious diseases, such as leishmaniasis (Azar & Nel 2003). Some Psychodidae are commonly nuisance pests in bathrooms, because their larvae (such as those from the subfamilies Psychodinae and Sycoracinae) live in aquatic or semi-terrestrial habitats including bathroom sinks (Grimaldi & Engel 2005).

Psychodid flies are well represented in dipteran fossil records (Meunier 1905; Quate 1961, 1963; Ansorge 1994, 1996; Evenhuis 1994; Azar et al. 1999, 2003, 2007a,b; Wagner 2001, 2002, 2006; Azar & Nel 2002, 2003; Nel et al. 2002; Azar & Ziadé 2005; Lak et al. 2008; Solórzano Kraemer & Wagner 2009; Azar & Waller 2010; Wang et al. 2011, among others). Their earliest incontestable known fossil is from the Lower Jurassic but their occurrence could be traced back to the Triassic (Ansorge 1994; Fraser et al. 1996; Blagoderov et al. 2007).

Mexican amber yield to date twelve fossil psychodid (Quate 1961, 1963; Evenhuis 1994) all from Oligocene-Miocene Simojovel deposits. We describe herein a new fossil psychodine fly from Totolapa (a recently rediscovered deposit). This fossil corresponds to the second species belonging to the fossil genus *Succinarisemus* Wagner, (2001) 2002, elderly known only from Dominican amber.

Totolapa is located in the central depression of Chiapas, 70 km at the south-east of Tuxtla Gutierrez, the capital of Chiapas State. The amber (with biological inclusions) comes from the Salt River mine (1 km north of Totolapa), exploited since 2007. The age of this amber is still polemic, supposed to be Oligocene-Miocene, but after the geological map of the Instituto Nacional de Estadística, Geografía y Informática (INEGI) (1985), the amber deposit would be located in Eocene sediments. Nevertheless two elements would rather support the Oligocene-Miocene age: (i) a ^{13}C-NMR spectroscopy study (Lambert et al. 1989) suggests the same palaeobotanical source for amber of Simojovel and Totolapa amber; (ii) the recent discovery of *Pheidole primigenia* Baroni Urbani, 1995, an ant from Totolapa that had been originally described from the Oligocene-Miocene Dominican amber. In any case, precise age is currently impossible to be given, but however we estimate that this amber, as for its possible Simojovel coeval, is probably Oligocene-Miocene.

Material and Methods

The amber piece was cut and shaped using a shaving blade, then final lustration was done using diatomite powder. The specimen was examined under a Nikon SZ10 stereomicroscope and a Leitz Laborlux 12 compound microscope, both equipped with drawing tubes for line drawings and Zeiss AxioCam ERc5s. The wing venational nomenclature of McAlpine (1981) is followed.

Systematic Paleontology

Family Psychodidae Newman, 1834
Subfamily Psychodinae Tonnoir, 1922
Genus *Succinarisemus* Wagner, (2001) 2002

Type species

Succinarisemus scheveni Wagner, (2001) 2002, by monotypy.

Succinarisemus totolapensis **sp. n.** (Figs 1–13)

Material

Holotype, specimen No. 104 (D. Coty collection), male, provisionally stored in the Muséum National d'Histoire Naturelle, Paris, France.

Age and horizon

Amber from Totolapa (Chiapas), Mexico, probably Oligocene-Miocene in age.

Figure 1. Photo of habitus of *Succinarisemus totolapensis* sp. n., holotype, specimen No. 104.

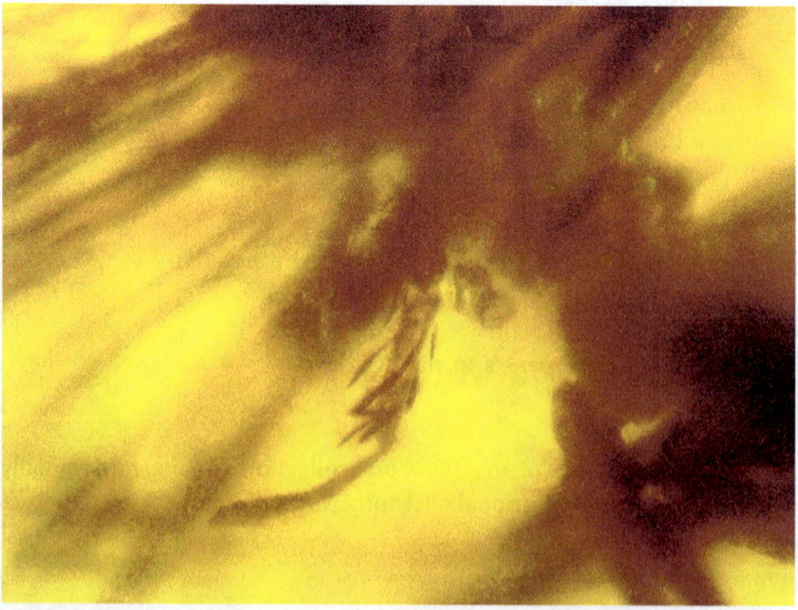

Figure 2. Photo of palpomeres of *S. totolapensis* sp. n., holotype, specimen No. 104.

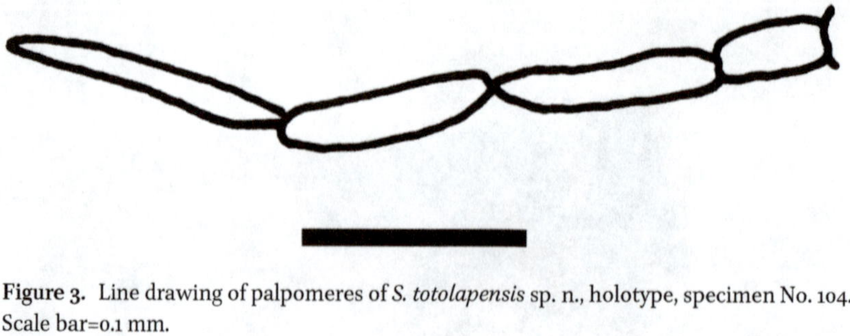

Figure 3. Line drawing of palpomeres of *S. totolapensis* sp. n., holotype, specimen No. 104. Scale bar=0.1 mm.

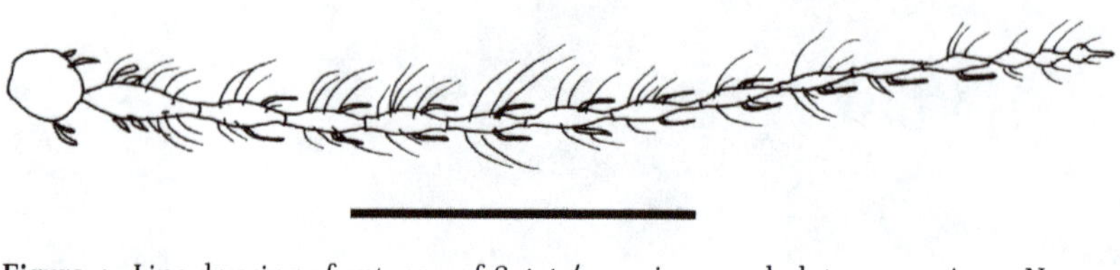

Figure 4. Line drawing of antenna of *S. totolapensis* sp. n., holotype, specimen No. 104. Scale bar=0.3 mm.

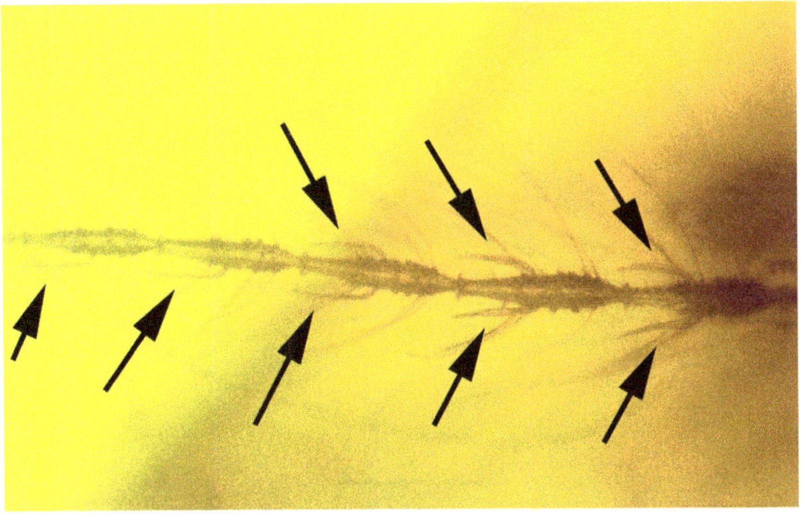

Figure 5. Photo of detail of antennomeres of *S. totolapensis* sp. n., holotype, specimen No. 104, arrows show ascoids.

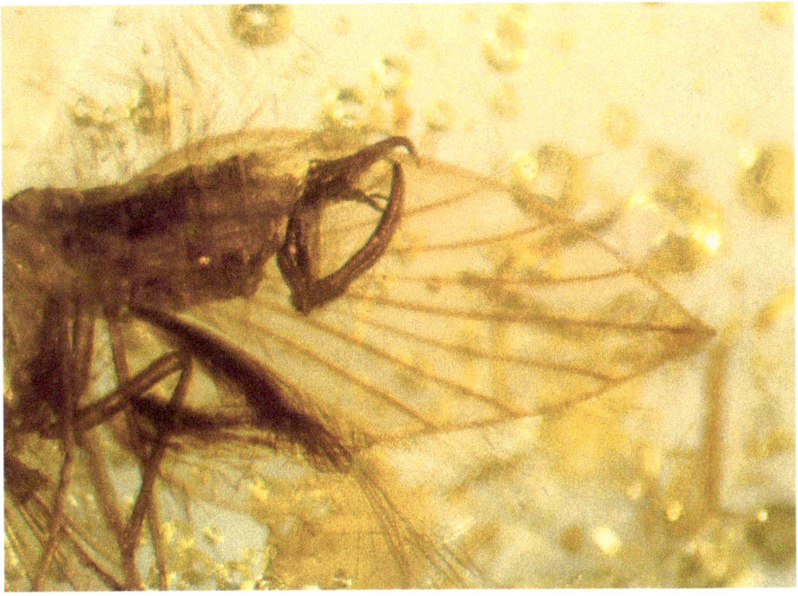

Figure 6. Photo of wing of *S. totolapensis* sp. n., holotype, specimen No. 104, arrows show ascoids.

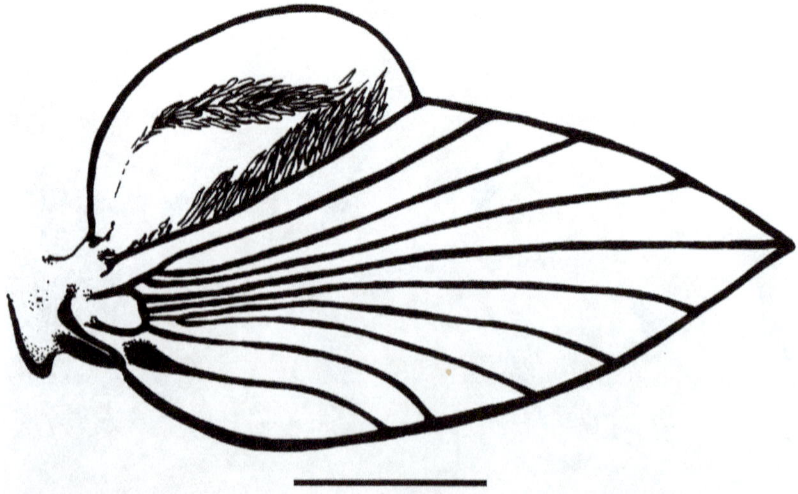

Figure 7. Line drawing of wing of *S. totolapensis* sp. n., holotype, specimen No. 104. Scale bar=0.5 mm.

Figure 8. Photo of male genitalia of *S. totolapensis* sp. n., holotype, specimen No. 104, lateral view.

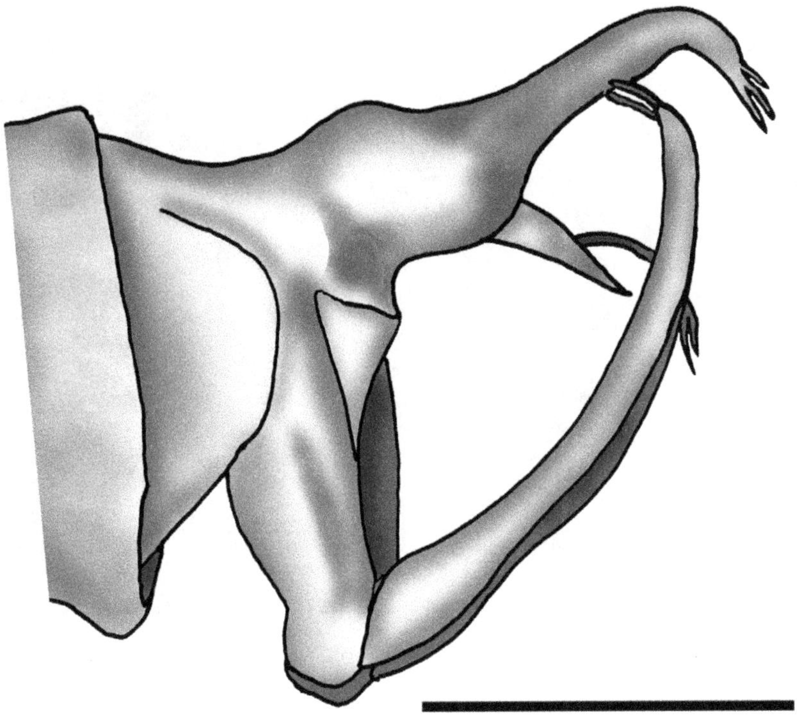

Figure 9. Line drawing of male genitalia of *S. totolapensis* sp. n., holotype, specimen No. 104, lateral view. Scale bar=0.3 mm.

Etymology

The species name is derived from Totolapa, the village where the amber deposit from which our specimen is originated was found.

Diagnosis

Wing area between costa and radius conspicuously broadened as in *Succinarisemus scheveni* Wagner, (2001) 2002. *Succinarisemus totolapensis* sp. n. differs from *S. scheveni* by its male genitalia structure: dististyles longer than basistyles, and bearing three strong distal setae instead of one; and in its incomplete eyebridge.

Description

Body 1.93 mm long. Head 0.42 mm wide, with incomplete eyebridge, 0.07 mm sparse. Palpus 4-segmented, basal segment short, 0.05 mm long,

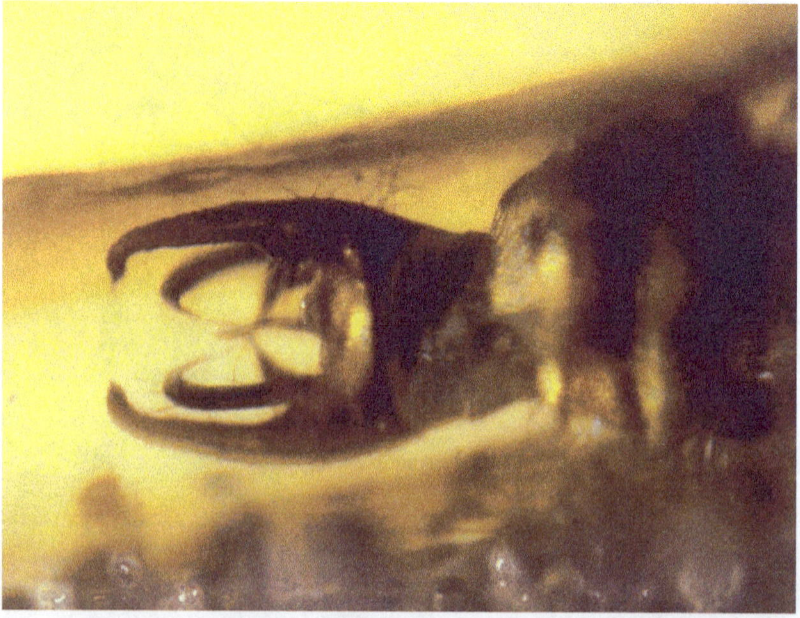

Figure 10. Photo of male genitalia of *S. totolapensis* sp. n., holotype, specimen No. 104, dorsal view.

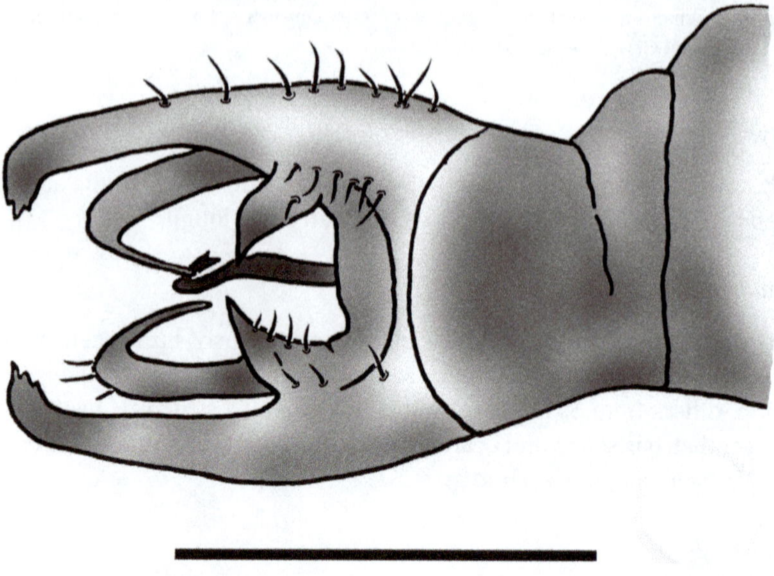

Figure 11. Line drawing of male genitalia of *S. totolapensis* sp. n., holotype, specimen No. 104, dorsal view. Scale bar=0.3 mm.

Figure 12. Photo of male genitalia of *S. totolapensis* sp. n., holotype, specimen No. 104, ventral view.

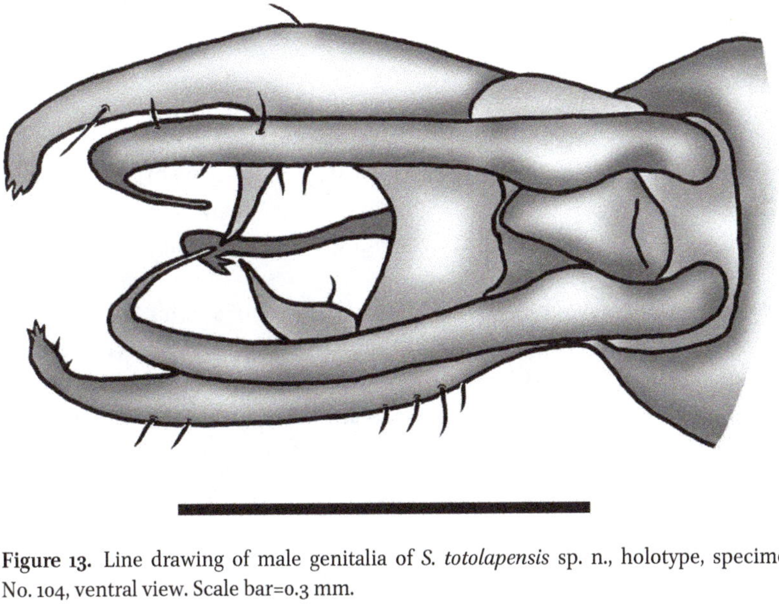

Figure 13. Line drawing of male genitalia of *S. totolapensis* sp. n., holotype, specimen No. 104, ventral view. Scale bar=0.3 mm.

segments 2 and 3 subequal, respectively 0.1 and 0.096 mm, distal segment elongate, 0.13 mm long, thinner than segments 2 and 3 (Figs 2 and 3). Antenna 1.01 mm long (Fig. 4), with 14 flagellomeres, fusiform, with digitiform ascoids (Fig. 5), paired on flagellomeres 2–8, and single on flagellomeres 9–11. Scape cylindrical 0.06 mm long and 0.09 mm wide, pedicel spherical about 0.06 mm long and wide. First flagellomere nearly 1.5 time longer than second one, flagellomeres 2–11 of nearly the same length but slightly decreasing apically, last three flagellomeres strongly reduced in size (Fig. 4). Thorax (0.5 mm) with translucent wing, setose, very slightly tinted, 2.09 mm long and 1.19 wide (Figs 6 and 7). Area between costa and radius conspicuously broadened. Wing venation with all forks basally. Neala triangular well developed and sclerotized. Knob 0.14 mm long and 0.07 mm wide. Stem 0.13 mm long. Legs short; mid and hind legs shorter than whole body. Foreleg 1.3 mm; forefemurae 0.3 mm; foretibiae 0.49 mm. Midleg 1.65 mm; midfemurae 0.51mm; midtibiae 0.58 mm; Hindleg 1.73 mm; hindfemurae 0.51 mm; hindtibiae 0.62 mm. Abdomen 0.78 mm long excluding genital appendages. Male genitalia (Figs 8–13) of inverted type. Basistyles conical, slightly bent internally, with a curved venterobasal strongly sclerotized and apically pointed appendage. It is not evident, whether basistyles and dististyles are linked with or without a hinge. Dististyles cylindrical, longer than basistyles, with three strong distal seta. Aedeagus visible, forked distally. Cercopodia very long, slightly curved, basally swollen internally, with a single short tenaculum.

Remarks

After Quate & Vockeroth (1981) and Wagner (1997), *Succinarisemus totolapensis* sp. n. falls in Psychodinae because of holding an eyebridge and fusiform flagellomeres. It belongs to *Succinarisemus* Wagner, (2001) 2002 as it has the same wing features with the autapomorphic enlarged wing area between costa and R1.

Succinarisemus totolapensis sp. n. presents a lot of similar features with *S. scheveni* Wagner (2001) 2002, but differs from the latter by its male genitalia structure (dististyles longer than basistyles, and bearing three strong distal setae instead of one); and in its incomplete eyebridge; which justify the creation of a new species.

Conclusion

Succinarisemus totolapensis sp. n. is the second species assigned to genus *Succinarisemus*. This latter was originally described on a *solo* fossil from

Dominican amber found in clastic sediments of an Oligocene to Middle Miocene age (Penney 2010). Solòrzano-Kraemer (2007, 2010) consider the Mexican amber (from Simojovel) as having the same age as the Dominican one but give the age of middle Miocene. It is true that our fossil comes from a different locality (Totolapa), but as indicated before we estimate that it is a possible coeval to Simojovel.

Wagner (2001) indicated that the conspicuously broadened wing area between costa and radius could be opened during courtship behaviour, and indicated that the radial vein could act as the frontal wing one in normal situation (as the right wing was folded on this level); such behaviour is known from an undescribed recent taxon from East Africa. In our fossil we have the same situation with one wing folded at the level of radial vein. It is likely that the statement of courtship performance is correct as the broadened wing area is covered with broad black setae, a common sexual dimorphic feature for males in psychodine flies. It is very regrettable that to date no females belonging to the genus *Succinarisemus* were found. The discovery of a species in the Mexican amber indicates that this genus was spread and living in the central America–Caribbean area during the Oligo-Miocene and increases our knowledge about palaeobiodiversity of psychodids.

Acknowledgements

We thank Manuel and Heriberto Ramirez, owners and miners of Salt River Mine, and their whole family, for their great help in our field research. This paper is a contribution to the project team: Biodiversity: Origin, Structure, Evolution, and Geology allotted to D.A. and Y.S. by the Lebanese University.

References

Ansorge, J. (1994) Tanyderidae and Psychodidae (Diptera) from the Lower Jurassic of Northeastern Germany. *Paläontologische Zeitschrift* **68**: 199–210.

Ansorge, J. (1996) *Eotonisca tertiaria* Meunier, 1905 (Psychodidae) aus dem Sachsischer Bernstein. *Studia Dipterologica* **3**: 195–199.

Azar, D. & Nel, A. (2002) New Cretaceous psychodid flies from Lebanese amber and Santana Formation (Chapada do Araripe, Brazil). (Diptera). *Annals de la Société Entomologique de France* **38**: 253–262.

Azar, D. & Nel, A. (2003) Fossil Psychodoid flies and their relation to parasitic diseases. *Mem. Inst. Oswaldo Cruz, Rio de Janeiro* **97** (Suppl. 1): 35–37.

Azar, D. & Waller, A. (2010). Redescription of *Eatonisca tertiaria* Meunier 1905 from Baltic amber (Diptera: Psychodidae: Trichomyiinae). *Annales de la Société Entomologique de France* **46**: 237–241.

Azar, D. & Ziadé, K. (2005) *Xenopsychoda harbi*, a new psychodoid fly from the Lower Cretaceous amber of Lebanon (Diptera; Psychodoidea). *Comptes Rendus Palevol.* 4: 25–30.

Azar, D., Nel, A., Solignac, M., Paicheler, J.-C. & Bouchet, F. (1999) New genera and species of phlebotomid and psychodid flies from the Lower Cretaceous amber of Lebanon (Insecta: Diptera: Phlebotomidae, Psychodidae). *Palaeontology* **42**: 1101–1136.

Azar, D., Perrichot, V., Néraudeau, D. & Nel, A. (2003) New psychodid flies from the Cretaceous ambers of Lebanon and France, with a discussion about *Eophlebotomus connectens* Cockerell, 1920 (Diptera, Psychodidae). *Annals of the Entomological Society of America* **96**: 117–127.

Azar, D., Adaymeh, C. & Jreich, N. (2007a) *Paleopsychoda zherikhini*, a new Cretaceous species of moth flies from Taimyr amber (Diptera: Psychodidae: Psychodinae). *African Invertebrates* **48**: 163–168.

Azar, D., Tahchy, A. & Perrichot, V. (2007b) The oldest Sycoracinae (Diptera: Psychodidae) from the French Cretaceous amber. *Alavesia* **1**: 5–10.

Baroni Urbani, C. (1995) Invasion and extinction in the West Indian ant fauna revisited: the example of Pheidole (Amber Collection Stuttgart: Hymenoptera, Formicidae. VIII: Myrmicinae, partim). *Stuttgarter Beiträge zur Naturkunde, Ser. B (Geol. Paläontol.)* **222**: 1–12.

Blagoderov, V.A., Grimaldi, D.A. & Fraser, N.C. (2007) How time flies for flies: diverse Diptera from the Triassic of Virginia and early radiation of the order. *American Museum Novitates* **3572**: 1–39.

Bryant, D.D. (1983) A recently discovered amber source near Totolapa, Chiapas, Mexico. *American Antiquity* **48**: 354–357.

Evenhuis, N.L. (1994) *Catalogue of the fossil flies of the World (Insecta: Diptera)*. Backhuys Publishers, Leiden, 600 pp.

Fraser, N.C., Grimaldi, D.A., Olsen, P.E. & Axsmith, B. (1996) A Triassic Lagerstätte from eastern North America. *Nature* **380**: 615–619.

Grimaldi, D. & Engel, M.S. (2005). *Evolution of the Insects*. Cambridge University Press, Cambridge, 755 pp.

INEGI (1985) *Carta Geológica, E15–11 (Tuxtla Gutiérrez), escala 1:250,000*. SPP/INEGI, Instituto Nacional de Estadística, Geografía y Informática, Mexico City.

Lambert, J.B., Frye, J.S., Lee Jr., T.A., Welch, C.J. & Poinar Jr., G.O. (1989). Analysis of Mexican amber by Carbon-13 NMR Spectroscopy. *Archaeological Chemistry* **4**: 381–388.

Lak, M., Azar, D., Nel, A., Néraudeau, D. & Tafforeau, P. (2008) The oldest representative of the Trichomyiinae (Diptera: Psychodidae) from the lower Cenomanian French amber studied with phase-contrast synchrotron X-ray imaging. *Invertebrate Systematics* **22**: 471–478.

McAlpine, J.F. (1981) Morphology and terminology – adult. In: McAlpine, J.F, Peterson, B.V., Shewell, G.E., Teskey, H.J., Vockeroth, J.R. & Wood, D.M. (Eds) *Monograph No. 27: Manual of Nearctic Diptera. Volume 1*. Research Branch Agriculture Canada, Ottawa, ON, pp. 9–63.

Meunier, F. (1905) Monographie des Psychodidae de l'ambre de la Baltique. *Annales Musei Nationalis Hungarici* **3**: 235–255.

Nel, A., Menier, J.-J. & De Ploëg, G. (2002) The oldest representative of the Trichomyiinae from the Lowermost Eocene amber of the Paris Basin (France) (Diptera: Psychodidae). *Annales de la Société Entomologique de France, (N.S.)* **38**: 247–252.

Newman, E. (1834) Attempted division of British insect into natural orders. *The Entomological Magazine* 2: 379–431.

Penney, D. (2010) Dominican amber. In: Penney, D. (Ed.) *Biodiversity of fossils in amber from the major world deposits.* Siri Scientific Press, Manchester, pp. 22–41.

Quate, L.W. (1961) Fossil Psychodidae in Mexican amber. Part 1. *Journal of Paleontology* 35: 949–951.

Quate, L.W. (1963) Fossil Psychodidae in Mexican amber. Diptera: Insecta. Part 2. *Journal of Paleontology* 37: 110–118.

Quate, L.W. & Vockeroth, J.R. (1981) Psychodidae. In: McAlpine, J.F, Peterson, B.V., Shewell, G.E., Teskey, H.J., Vockeroth, J.R. & Wood, D.M. (Eds) *Monograph No. 27: Manual of Nearctic Diptera. Volume 1.* Research Branch Agriculture Canada, Ottawa, ON, pp. 293–300.

Solórzano-Kraemer, M.M. (2007) Systematic, palaeoecology, and palaeobiogeography of the insect fauna from Mexican amber. *Palaeontographica* (A) 282: 1–133.

Solórzano-Kraemer, M.M. (2010) Mexican amber. In: Penney, D. (Ed.) *Biodiversity of fossils in amber from the major world deposits.* Siri Scientific Press, Manchester, pp. 42–56.

Solórzano-Kraemer, M.M. & Wagner, R. (2009) The first psychodid (Diptera: Psychodidae: Phlebotominae) species from the Lower Eocene amber of Vastan, Gujarat, India. *Zootaxa* 2152: 63–68.

Tonnoir, A.L. (1922) Note sur le genre Nemopalpus (Psychodidae) et description d'une espèce nouvelle. *Annales de la Société Entomologique de Belgique* 62: 125–136.

Wagner, R. (1997) Family Psychodidae. In: Papp, L. & Darvas, B. (eds.), *Contributions to a Manual of Palaearctic Diptera. Vol. 2: Nematocera and Lower Brachycera.* Science Herald, Budapest, pp. 205–226.

Wagner, R. (2001) A remarkable new species and genus of moth-flies (Diptera, Psychodidae, Psychodinae) from Dominican amber. *Studia Dipterologica* 8: 423–426.

Wagner, R. (2002). *Succinarisemus scheveni* spec. nov. et gen. nov. (Diptera, Psychodidae, Psychodinae) from Carribean amber - Validation. *Studia Dipterologica* 9: 30.

Wagner, R. (2006) Amber Bruchomyiinae - descriptions of already known and new species, and the position of the 'subfamily' within Psychodidae (s.l.) (Diptera). *Studia Dipterologica* 13: 83–95.

Wang, B., Zhang, H., and Azar, D. (2011) The first Psychodidae (Insecta: Diptera) from the lower Eocene Fushun amber of China. *Journal of Paleontology* 85: 1154–1159.

A new *Sycorax* species from Eocene Ukrainian Rovno Amber (Diptera: Psychodida: Sycoracinae)

Dany Azar[a], André Nel[b] and Evgeny Perkovsky[c]

[a]Lebanese University, Faculty of Sciences II,
Department of Natural Sciences, P.O. Box 26110217, Fanar – Matn, Lebanon
E-mail: azar@mnhn.fr
[b]CNRS UMR 7205, Muséum National d'Histoire Naturelle, CP 50, Entomologie,
45 rue Buffon, F-75005 Paris, France
E-mail: anel@mnhn.fr
[c]Schmalhausen Institute of Zoology, Bogdan Khmelnitski Street 15, Kiev 01601, Ukraine
E-mail: perkovsky@fromru.com

Abstract

A new sycoracine species is described and figured from male and female specimens from Late Eocene Rovno amber (Ukraine). *Sycorax ukrainensis* sp. n. is the first psychodid fly described from this material though this later includes several species common in Baltic amber. The new species adds evidence that sycoracine were once more diverse and more widespread and global, in the past. The new species is distinguished from related taxa, and a discussion is provided

Keywords

Psychodidae, Sycoracinae, sp. n., Eocene, Ukraine

Introduction

Sycoracinae are tiny flies with some species bearing functional mandibles and having blood-feeding behaviour like phlebotomine flies. They constitute a small group within Psychodidae in modern fauna (less than 40 described species). They are currently represented by three genera, i.e., *Sycorax* Haliday *in* Curtis, 1839, *Aposycorax* Duckhouse, 1972, and *Parasycorax* Duckhouse, 1972, even if some authors consider that all members of this subfamily should be placed in a single genus *Sycorax* (Ježek 1999; Bejarano et al. 2008).

The Sycoracinae were first classified within the subfamily Trichomyiinae after the wing vein Rs three-branched with four radial veins instead of Rs four-branched and with five radial veins as in the other psychodids

(R4 and R5 being fused and forming R4+5 in both Sycoracinae and Trichomyiinae). Edwards (1929) recognized the difference between *Sycorax* and *Trichomyia* Haliday, 1839 (Sycoracinae were only represented by that time by the genus *Sycorax*), and suggested that *Sycorax* should be classified in a separate tribe of the Trichomyiinae or preferably as a separate subfamily. Jung (1954, 1956) recognised a separate subfamily for *Sycorax*, but later he reverted to the single subfamily Trichomyiinae (Jung 1958), like Hennig (1972). Duckhouse (1972), Vaillant (1978) and Wagner (1997) considered Sycoracinae as a separate subfamily. The fossil record of this subfamily is relatively rich with the oldest representative *Sycorax neli* Azar, Tahchy & Perrichot, 2007 from the Lower Cenomanian amber of Cadeuil, Charente-Maritime (SW France) (Azar et al. 2007). The Sycoracinae appears as an ancient group within psychodoids, apparently much more diversified in the past than in the Recent entomofauna.

Herein we describe a new fossil species of psychodid fly belonging to the genus *Sycorax* from the Late Eocene Rovno amber, Ukraine. This fossil was totally unexpected as the psychodid fauna of Rovno amber has a broad similarity with the Baltic amber one.

Material and Methods

The fossils in amber were carefully prepared in Canada balsam medium, following the method described by Azar et al. (2003), in order to observe as many characters as possible and to preserve the material from 'weathering'. Thus the 'absences' of structures are accurate, which is different of structures that are 'not visible' but maybe present. The specimen was examined with an Olympus SZX9 stereomicroscope and an Olympus inverted compound microscope CK40, both equipped with a camera lucida for line drawings and digital cameras. For the description of the vein nomenclature of the wing we follow McAlpine (1981).

Systematic Palaeontology

Order Diptera Linnaeus, 1758
Family Psychodidae Newman, 1834
Subfamily Sycoracinae Jung, 1954
Genus *Sycorax* Haliday *in* Curtis, 1839

Type species

Sycorax silacea Haliday *in* Curtis, 1839, pp. 745.

***Sycorax ukrainensis* sp. n.** (Figs 1–19)

Etymology

The species name is after Ukraine, from which the amber subject of this study is originated.

Type material

Holotype. Specimen No. K-26806 (male) (Fig. 1).
Allotype. Specimen No. K-4027 (female) (Fig. 2); both deposited in the amber collection of the Schmalhausen Institute of Zoology NAN of Ukraine (SIZK), Kiev, Ukraine.

Age and locality

Late Eocene, Rovno Amber, Klesov (Perkovsky et al. 2010).

Diagnosis

Typical *Sycorax*-like wing venation, with R1 reaching C beyond radial fork; this fork apical to median one; CuA2 and A reaching posterior margin. Head with well developed mouthparts; palps with four palpomeres of

Figure 1. *Sycorax ukrainensis* sp. n., holotype specimen No. K-26806 (male), photograph of general habitus.

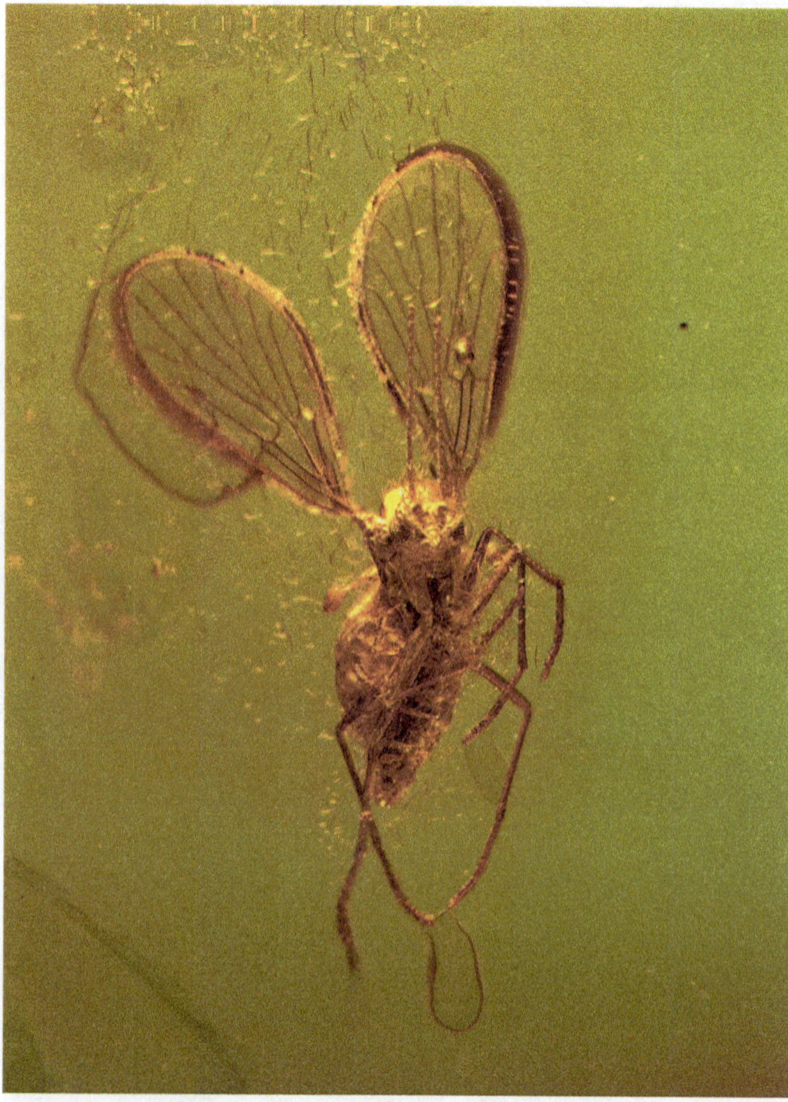

Figure 2. *Sycorax ukrainensis* sp. n., allotype specimen No. K-4027 (female), photograph of general habitus.

nearly equal size; second palpomere with sensorial pit in female; antenna with 13 flagellomeres; first one nearly 1.5 times as long as second; flagellomeres 2–5 of equal length; last one the wider and bearing a reduced and drop-like apiculus. Male genitalia with coxite and stylus simple and nearly equal in size, the latter bearing a powerful terminal spine.

Description

All measurements in mm, they were made on the holotype except when the contrary is indicated. Head, 0.3 long, 0.32 wide (on allotype), with slightly elliptical eyes (0.14 of largest diameter) widely separated, about 0.12 apart (Figs 3 and 4). Well developed vertex somewhat puffed out behind antenna. Antenna setose, 0.8 long (0.85 in allotype). First antennomere (scapus) subcylindrical, nearly 0.4 long and 0.35 wide; second antennomere (pedicel) subglobular, about 0.5 long and 0.5 wide. 13 flagellomeres; the first nearly 1.5 times as long as second, 0.9 long; flagellomeres 2–5 of equal length (0.7 each); then flagellomeres apically decreasing progressively in length; last one the wider and bearing a reduced and

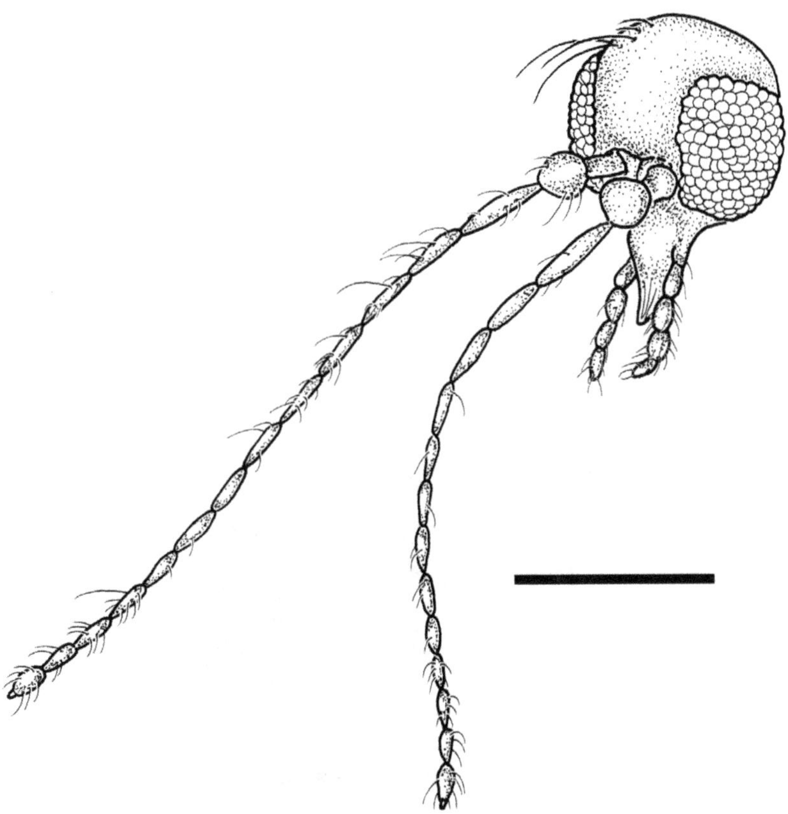

Figure 3. *Sycorax ukrainensis* sp. n., holotype specimen No. K-26806 (male), drawing of head. Scale bar=0.2 mm.

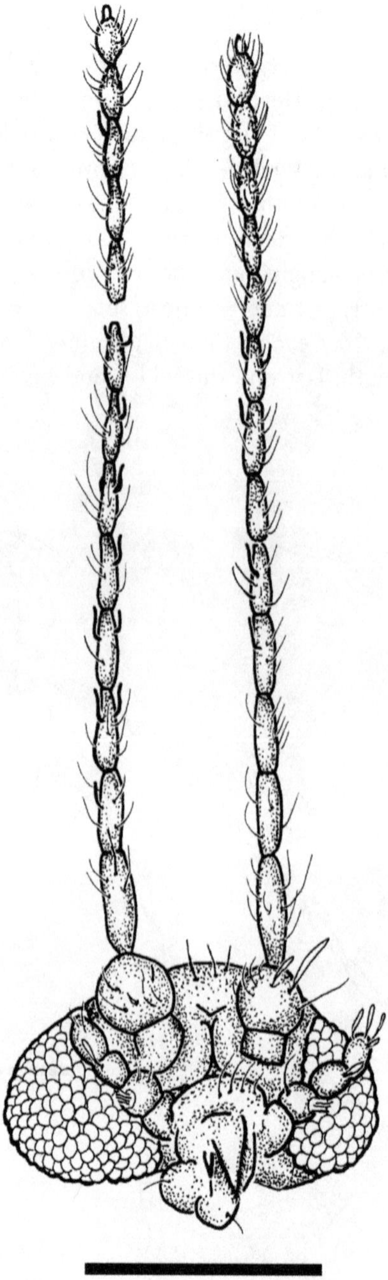

Figure 4. *Sycorax ukrainensis* sp. n., allotype specimen No. K-4027 (female), drawing of head. Scale bar=0.2 mm.

drop-like apiculus (Figs 1 and 2). Flagellomeres (observed on allotype) with a pair of digitate and sigmoid ascoids in apical third (Fig. 5). Palpus 0.12 long, with four hairy palpomeres of nearly equal lengths, 0.3 long (Figs 6–9). In female, second palpomere with sensorial pit carrying specialized rods (Figs 6–8); and third palpomere bearing a specialized sensorium apically (Fig. 6). Mouthparts well developed in both sexes, 0.12 long, with serration, mandibles toothed as in Figs 10 and 11. Thorax nearly 0.40 long, 0.35 wide (allotype). Pronotum gibbous. Legs long, longer than whole body. Wing 1.31 long, 0.55 wide (1.5 and 0.6 respectively in female), hyaline (Figs 12–15). Humeral vein (h) reaching costal margin at 0.16 from wing base. Subcostal vein Sc distally fused with R1 in a strong angle, 0.51 from wing base, and with a crossvein reaching costal margin. R1 reaching costal margin 0.99 from wing base. Rs separated from R1 0.47 from wing base, at nearly same level of fork of M into M1+2 and M3. Rs three-branched, with all its branches extending to wing margin. R2 and R3 separated 0.93 distally. R4 and R5 fused forming R4+5, 0.65 long, very slightly curved at its middle toward posterior side. Crossvein r-m 0.67 distal of wing base.

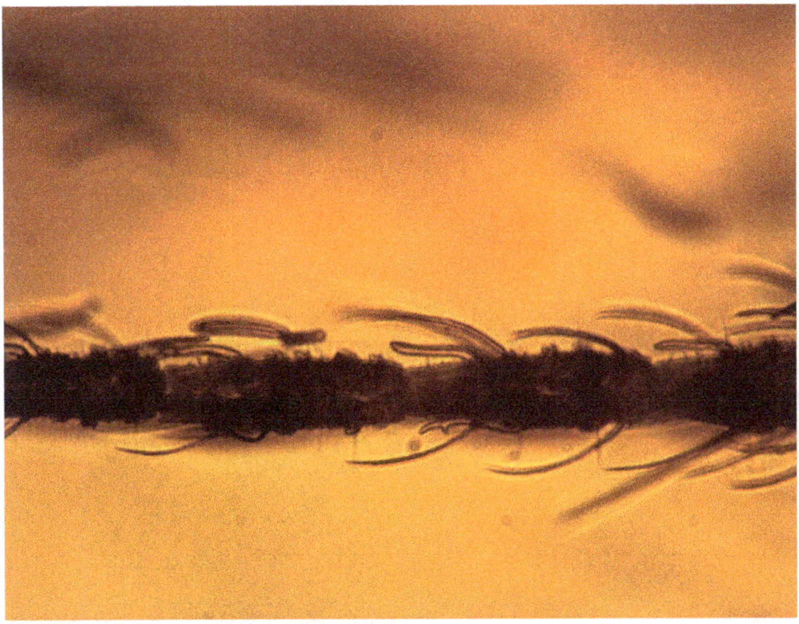

Figure 5. *Sycorax ukrainensis* sp. n., allotype specimen No. K-4027 (female), photograph of antennal ascoids.

Figure 6. *Sycorax ukrainensis* sp. n., allotype specimen No. K-4027 (female), photograph of lateral view of palp showing the sensorial pit on second palpomere (arrow).

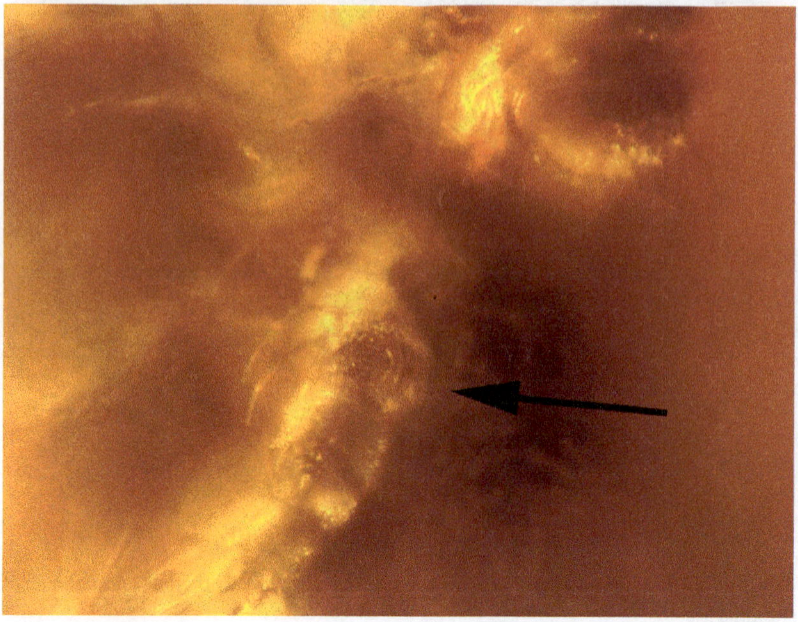

Figure 7. *Sycorax ukrainensis* sp. n., allotype specimen No. K-4027 (female), photograph of frontal view of palp showing the sensorial pit on second palpomere (arrow).

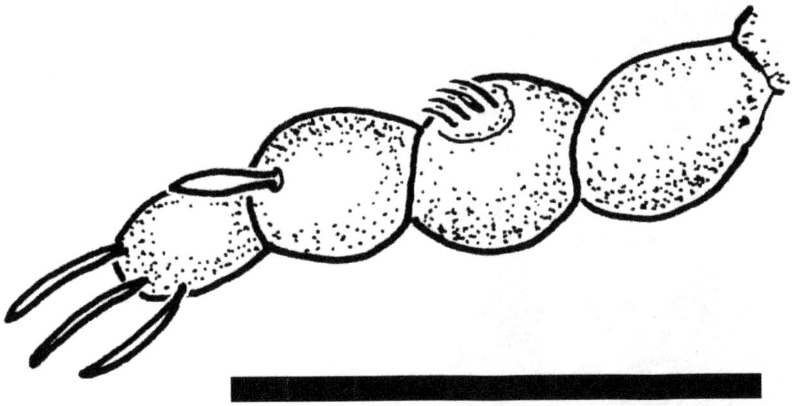

Fig. 8. *Sycorax ukrainensis* sp. n., allotype specimen No. K-4027 (female), drawing of palp. Scale bar=0.1 mm.

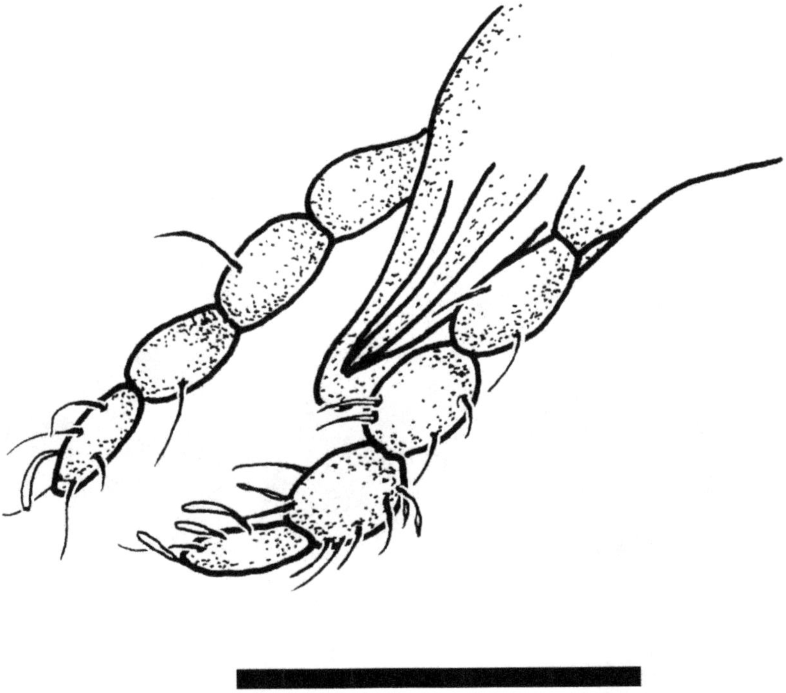

Figure 9. *Sycorax ukrainensis* sp. n., holotype specimen No. K-26806 (male), drawing of mouthparts plus palps. Scale bar=0.1 mm.

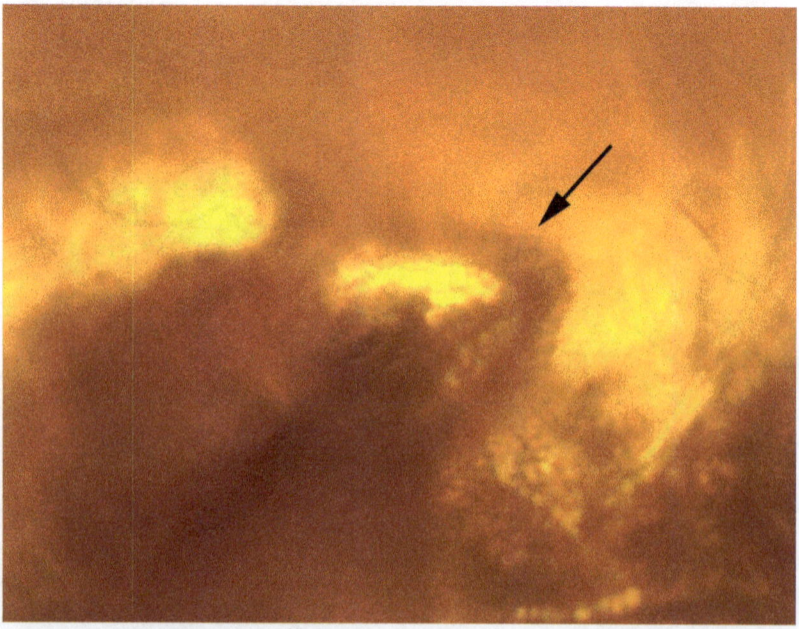

Figure 10. *Sycorax ukrainensis* sp. n., allotype specimen No. K-4027 (female), photograph of apex of lacinia showing serrations (arrow).

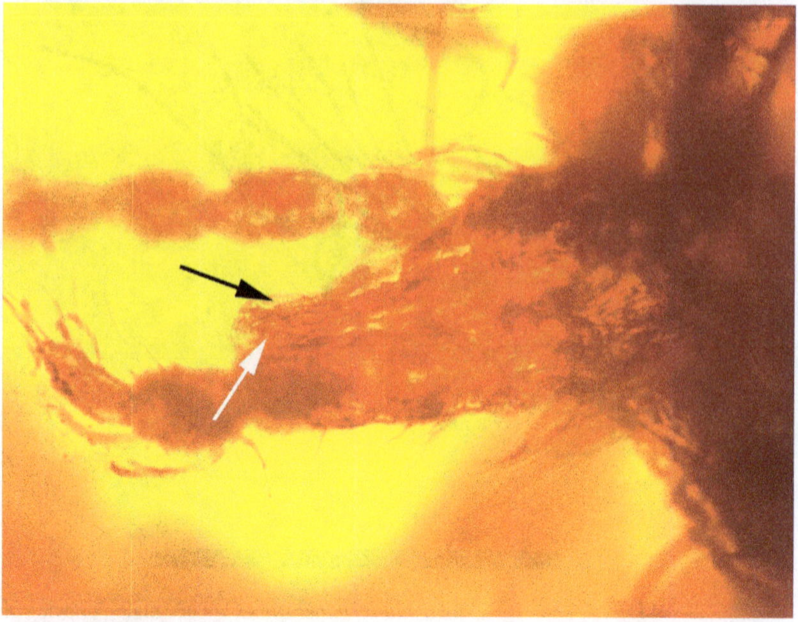

Figure 11. *Sycorax ukrainensis* sp. n., holotype specimen No. K-26806 (male), photograph of mouthparts showing toothed mandibles (white arrow) and lacinia (black arrow).

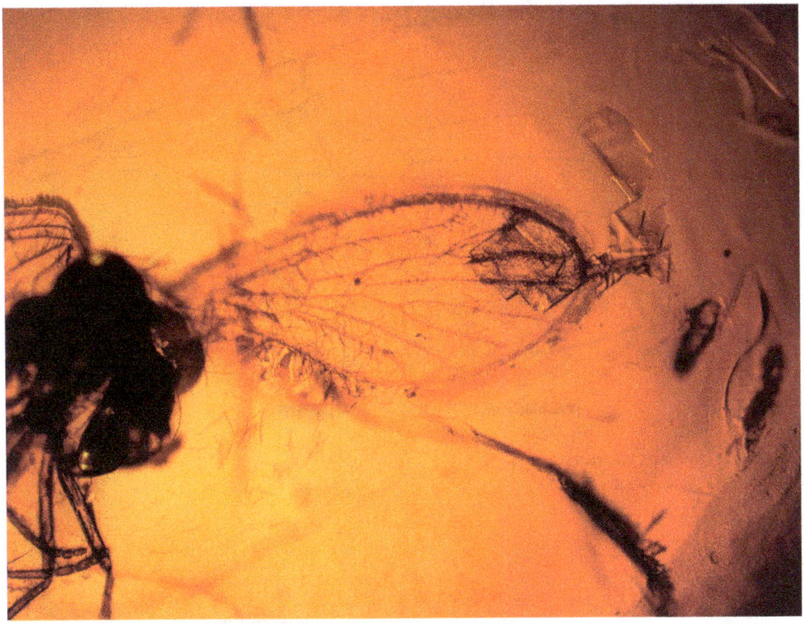

Figure 12. *Sycorax ukrainensis* sp. n., holotype specimen No. K-26806 (male), photograph of wing.

Figure 13. *Sycorax ukrainensis* sp. n., allotype specimen No. K-4027 (female), photograph of wing.

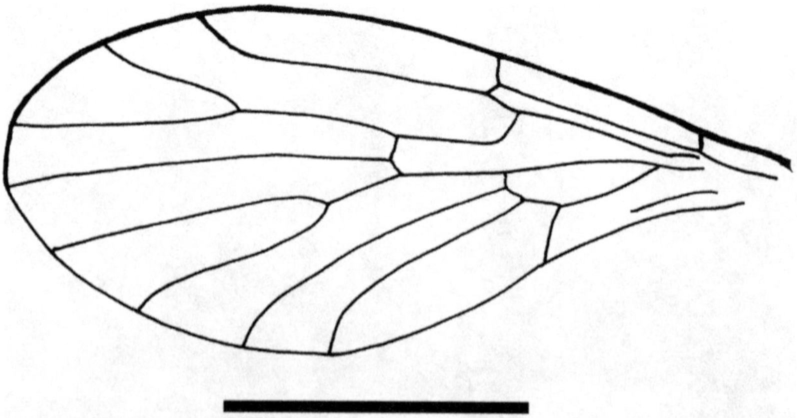

Figure 14. *Sycorax ukrainensis* sp. n., holotype specimen No. K-26806 (male), drawing of wing. Scale bar=0.5 mm.

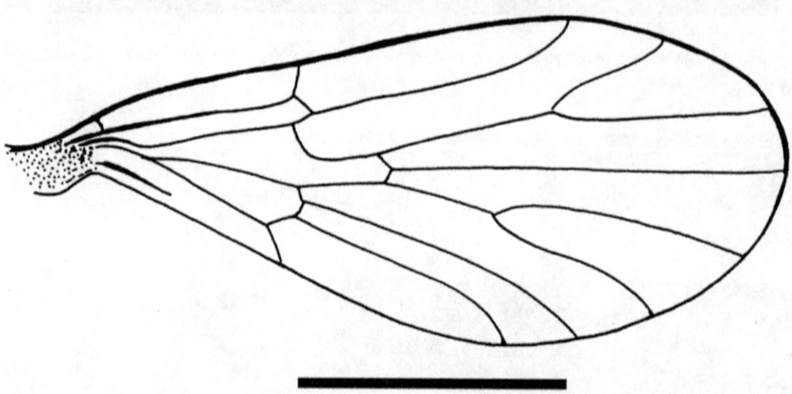

Figure 15. *Sycorax ukrainensis* sp. n., allotype specimen No. K-4027 (female), drawing of wing. Scale bar=0.5 mm.

Fork of M into M1+2 and M3 0.76 distal of wing base. Fork of M1+2 into M1 and M2 0.30 distal of base of M1+2. M1 slightly curved. M2 slightly sigmoidal, slightly shorter than M1. M3 reaching wing margin at 0.88 from wing base. CuA1 separating from CuA 0.38 distal of wing base reaching wing margin at 0.74 apically. CuA2 rather short, 0.10 long, reaching wing margin at 0.40 from wing base. A1 very short, not reaching posterior wing margin. Halteres of female 0.28 long; with knob 0.12 long; and stem 0.16 long. Legs

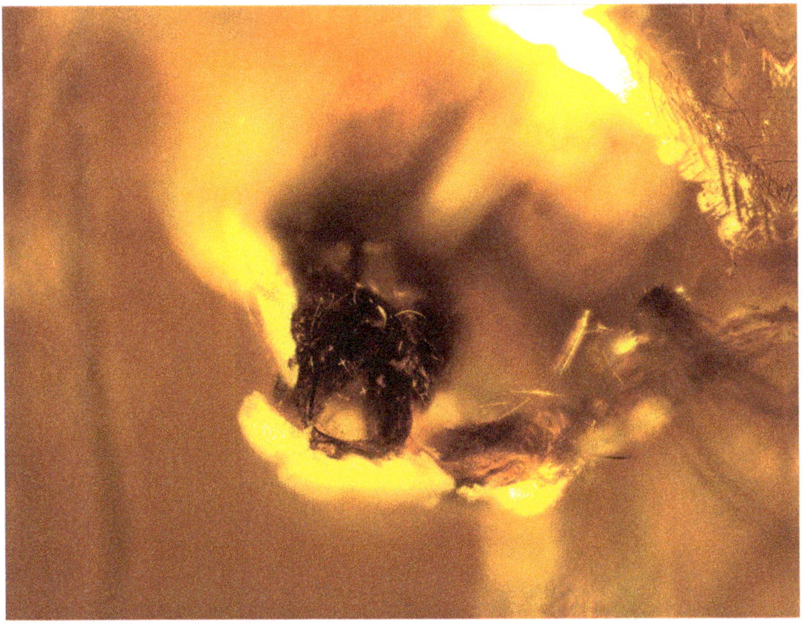

Figure 16. *Sycorax ukrainensis* sp. n., holotype specimen No. K-26806 (male), photograph of genitalia.

robust, slightly longer than body. Abdomen 0.85 long excluding genital appendages, 0.50 wide (female allotype). Dorsal surfaces of all abdominal segments bearing few setae. Male genital appendages (Figs 16 and 17) of not inverted type as for most of the Sycoracinae. Coxite and stylus simple, nearly cylindrical and equal in size. Coxite 0.10 long and 0.04 wide with a long setae on apical third. Stylus 0.09 long, 0.02 wide; bearing one apical powerful spine, 0.06 long. Cercopods and aedeagus as shown in Figs 16 and 17. Female genital appendages with two small subgenital plates and two elliptic cerci long (Figs 18 and 19).

Discussion

Male genitalia are not inverted in all the representatives of the genus *Sycorax*, but not in *Aposycorax chilensis* (Tonnoir, 1929) and in the *Parasycorax* species. In addition *Parasycorax* species have a stylus bearing three sharp spines instead of one in *Sycorax*. *Sycorax ukrainensis* sp. n. has male genitalia of not inverted type and with a stylus bearing one apical

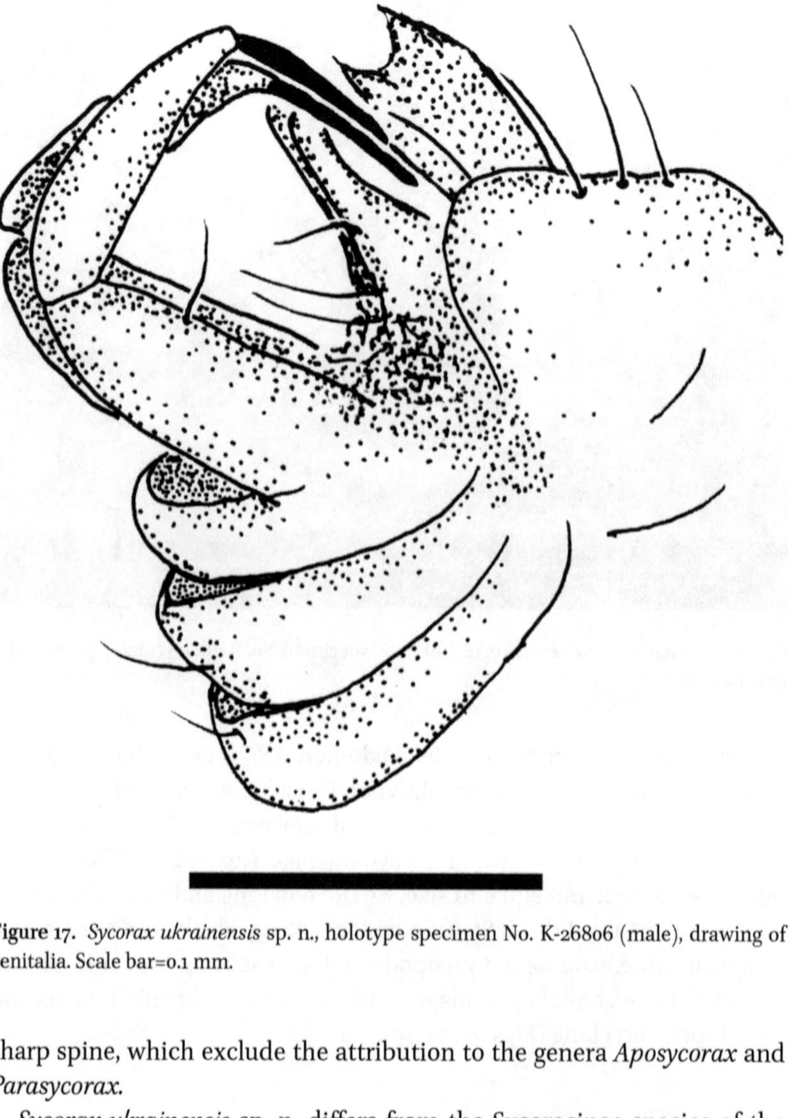

Figure 17. *Sycorax ukrainensis* sp. n., holotype specimen No. K-26806 (male), drawing of genitalia. Scale bar=0.1 mm.

sharp spine, which exclude the attribution to the genera *Aposycorax* and *Parasycorax*.

Sycorax ukrainensis sp. n. differs from the Sycoracinae species of the Baltic amber, currently considered as a northern coeval analogue of the Rovno Amber (Perkovsky et al. 2007, 2010) as follows: *Sycorax tumultuosa* Meunier, 1905 has the radial fork longer than median one, and the fourth palpomere as long as the three preceding altogether; whereas in our fossil the radial fork is shorter than median one and all the palpomere have nearly the same length; *Sycorax prompta* Meunier, 1905 has a lot of similarities with *Sycorax ukrainensis* sp. n. in the wing venation but differs in

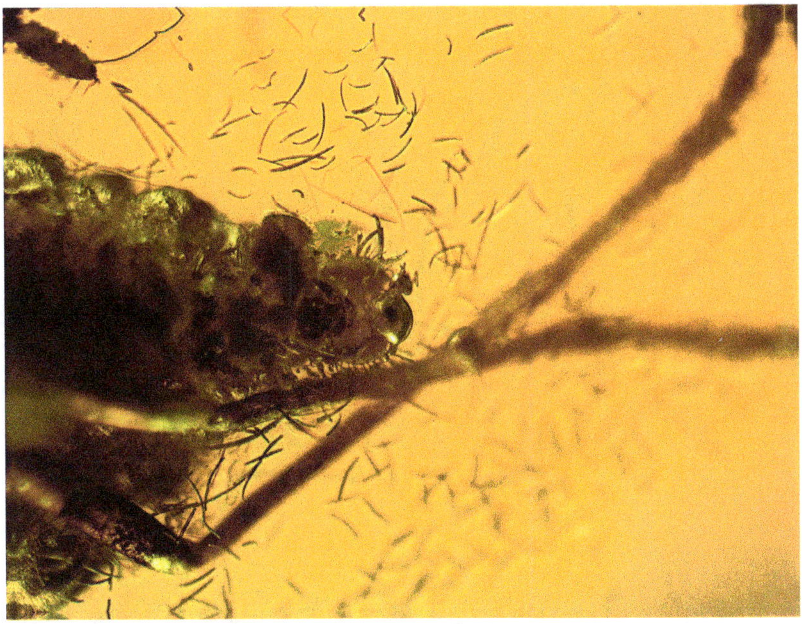

Figure 18. *Sycorax ukrainensis* sp. n., allotype specimen No. K-4027 (female), photograph of ventral view of apex of abdomen.

its shorter antenna and its much longer first flagellomere. *Sycorax ukrainensis* differs from *Sycorax peruensis* Petrulevičius et al., 2011 (from Peruvian amber) by having 13 flagellomeres instead of 12 (Petrulevičius et al. 2011). It differs from *Sycorax neli* Azar, Tahchi & Perrichot, 2007 (from the Lower Cenomanian amber of Cadeuil, Charente-Maritime, SW France) by having CuA1 and CuA2 ending in C and the first flagellomere longer than second one.

Note

A third *Sycorax* specimen is present in the same collection from Rovno amber (female specimen No. K-5982) (Figs 20–23). This fossil presents a lot of similarities with *Sycorax ukrainensis* but it differs from it in the more slender and shorter antenna, with a first flagellomere very long, nearly twice as long as second, CuA1 not touching posterior wing margin, and first palpomere the shorter, second slightly longer, third slightly longer than second, and fourth nearly as long as third. This fossil better corresponds to *Sycorax prompta*, especially to the "variété" with the first flagellomere distinctly longer than the second (male No. 2276 and female

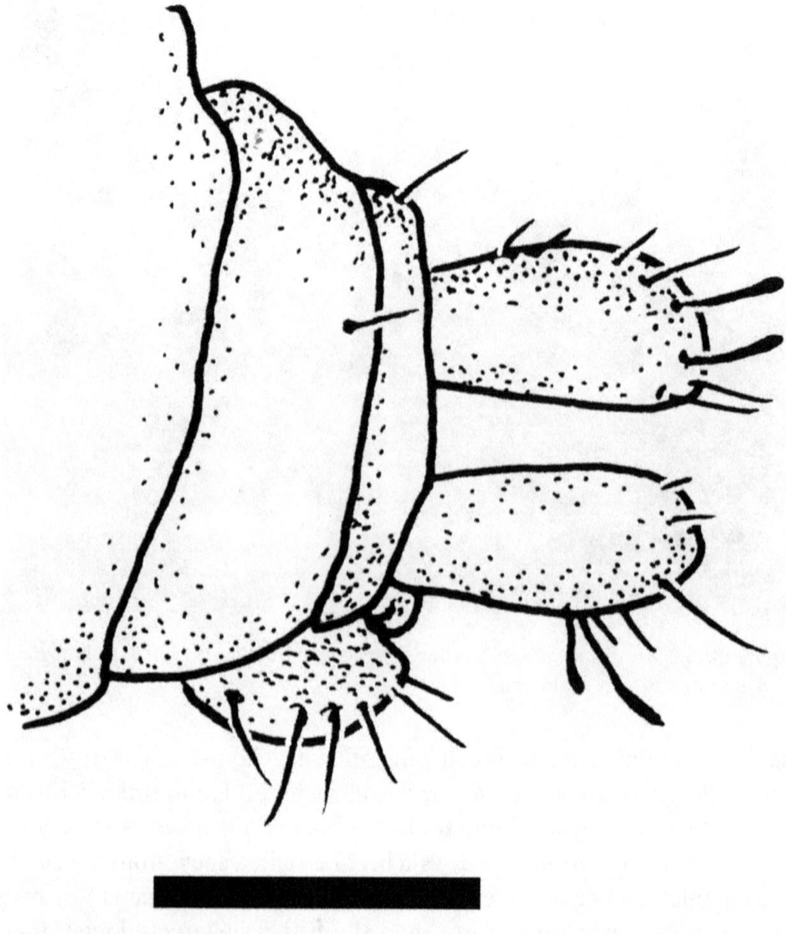

Figure 19. *Sycorax ukrainensis* sp. n., allotype specimen No. K-4027 (female), drawing of female genitalia. Scale bar=0.1 mm.

No. 7977 in the original description of Meunier) (Meunier 1905: 252, pl. 7, fig. 9). The relative dimensions of the palpomeres of No. K-5982 also correspond to those indicated for male No. 2276 by Meunier. It remains that the type (specimen No. 7734) of *S. prompta* and another specimen No. 7788 have different antennae with first flagellomere only slightly longer than second. Meunier also indicated the presence of an 'intermediate' form (specimen No. 2539) with the first flagellomere only twice as long as second. This confusing situation suggests the presence of at least two different species in the type series of *S. prompta*, but only the revision

Figure 20. *Sycorax prompta* Meunier, 1905, female specimen No. K-5982, photograph of general habitus.

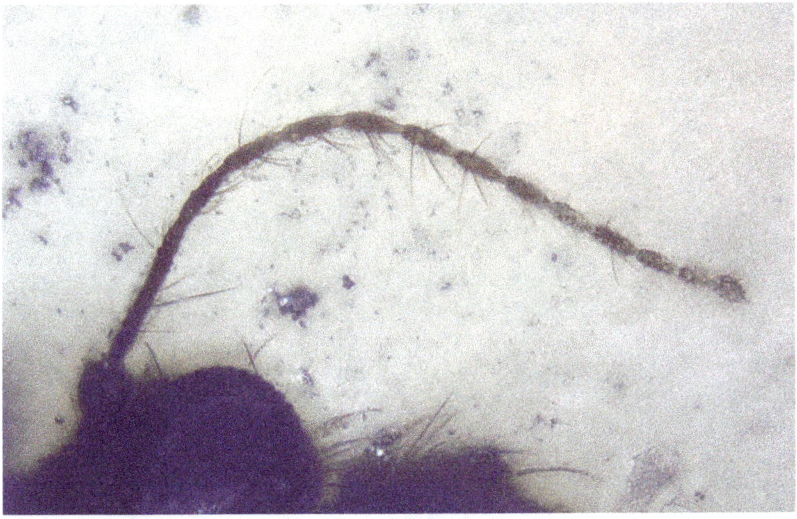

Figure 21. *Sycorax prompta* Meunier, 1905, female specimen No. K-5982, photograph of antenna.

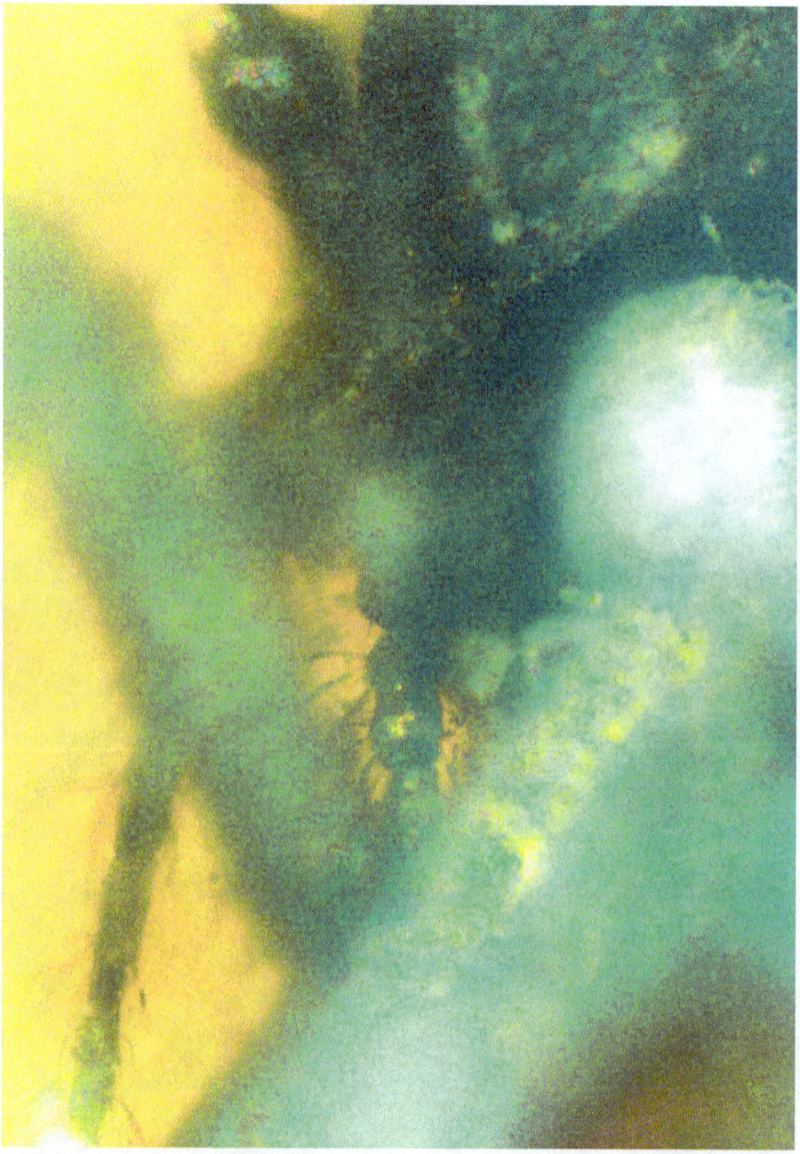

Figure 22. *Sycorax prompta* Meunier, 1905, female specimen No. K-5982, photograph of palp.

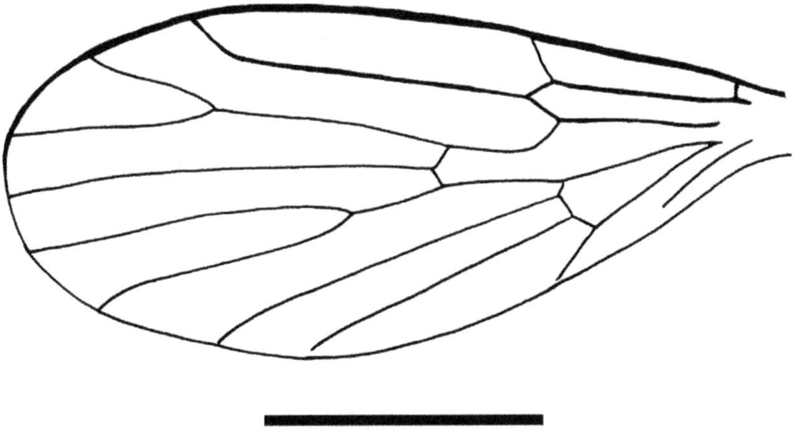

Figure 23. *Sycorax prompta* Meunier, 1905, female specimen No. K-5982, drawing of wing. Scale bar=0.5 mm.

of this material would solve the problem. We provisionally attribute specimen No. K-5982 to *Sycorax prompta*.

Acknowledgements

We are thankful to unknown reviewers for comments on a previous version. This paper is a contribution to the team project "Biodiversity: Origin, Structure, Evolution and Geology" awarded to D.A. by the Lebanese University.

References

Azar, D., Perrichot, V., Néraudeau, D. & Nel, A. (2003) New psychodid flies from the Cretaceous ambers of Lebanon and France, with a discussion about *Eophlebotomus connectens* Cockerell, 1920 (Diptera, Psychodidae). *Annals of the Entomological Society of America* **96**: 117–127.

Azar, D., Tahchy, A. & Perrichot, V. (2007) The oldest Sycoracinae (Diptera: Psychodidae) from the French Cretaceous amber. *Alavesia* **1**: 5–10.

Bejarano, E.E., Duque, P. & Velez, I.D. (2008) A new species of *Sycorax* (Diptera, Psychodidae, Sycoracinae) from the Pacific coast of Colombia. *Revista Brasileira de Entomologia* **52**: 28–31.

Edwards, F.W. (1929) A note on the amber moth-fly *Eophlebotomus connectens* Cockerell. *Annals and Magazine of Natural History* **10**: 424–425.

Duckhouse, D.A. (1972) Psychodidae (Diptera, Nematocera) of South Chile, subfamilies Sycoracinae and Trichomyiinae. *Transactions of the Royal Entomological Society* **124**: 231–268.

Hennig, W. (1972) Insektenfossilien aus der unteren Kreide. IV. Psychodidae (Phlebotominae), mit einer kritischen Übersicht über das phylogenetische System der familie und die bisher beschriebenen Fossilien (Diptera). *Stuttgarter Beiträge zur Naturkunde (B)* **241**: 1–69.

Ježek, J. (1999) Comments on the correct grammatic gender of *Sycorax* Curt. and *Philosepedon* Eat. (Diptera: Psychodidae) with world catalogue. *Dipterologica Bohemoslovaca* **9**: 83–87.

Jung, H.F. (1954) Einige neue mitteleuropäische Psychodiden (Diptera). *Zoologischer Anzeiger* **152**: 16–31.

Jung, H.F. (1956) Beitrage zur Biologie, Morphologie und Systematik der europaischen Psychodiden (Diptera). *Deutsche Entomologische Zeitschrift* (N.F.) **3**: 97–257.

Jung, H.F. (1958) Psychodidae. Chapter 9b. Trichomyiinae. In: Lindner, E. (Ed.) *Der Fliegen der Palaearktischen Region, Volume 3*. E. Schweizerbart'sche Verlagsbuchhandlung, Stuttgart, pp. 1–16.

McAlpine, J.F. (1981) Morphology and terminology – adult. In: McAlpine, J.F., Peterson, B.V., Shewell, G.E., Teskey, H.J., Vockeroth, J.R. & Wood, D.M. (Eds) *Manual of Nearctic Diptera. Volume 1. Research Branch Agriculture Canada Monograph Volume 27*. Research Branch Agriculture Canada, Ottawa, ON, pp. 9–63.

Meunier, F. (1905) Monographie des Psychodidae de l'ambre de la Baltique. *Annales Historico-Naturales Musei Nationalis Hungarici* **3**: 235–255.

Perkovsky, E.E., Rasnitsyn, A.P., Vlaskin, A.P. & Taraschuk, M.V. (2007) A comparative analysis of the Baltic and Rovno amber arthropod faunas: representative samples. *African Invertebrates* **48**: 229–245.

Perkovsky, E.E., Zosimovich, V.Yu. & Vlaskin, A.P. (2010). Rovno amber. In: Penney, D. (Ed.) *Biodiversity of fossils in amber from the major world deposits*. Siri Scientific Press, Manchester, pp. 116–136.

Petrulevičius, J.F., Nel, A., De Franceschi, D., Goillot, C., Antoine, P.-O., Salas-Gismondi, R. & Flynn, J.J. (2011) First fossil blood sucking Psychodidae in South America: a sycoracine moth fly (Insecta: Diptera) in the middle Miocene Amazonian amber. *Insect Systematics & Evolution* **42**: 87–96.

Tonnoir, A.L. (1929) *Diptera of Patagonia and south Chile. Part II, Fascicle I, Psychodidae*. British Museum of Natural History, London, 32 pp.

Vaillant, F. (1978) Contribution à l'étude des *Sycorax* (Dipt. Psychodidae) de la France. *L'Entomologiste* **34**: 70–77.

Wagner, R. (1997) Family Psychodidae. In: Papp, L. & Darvas, B. (Eds) *Contributions to a Manual of Palaearctic Diptera. Volume 2: Nematocera and Lower Brachycera*. Science Herald, Budapest, pp. 205–226.

Another new representative of the isometopine genus *Archemyiomma* (Hemiptera: Heteroptera: Miridae) from Late Eocene Rovno (Ukraine) amber

Aleksander Herczek[a], Yuri A. Popov[b] and Evgeny E. Perkovsky[c]

[a]Silesian University, Department of Zoology, Bankowa 9, 40 Katowice, Poland
e-mail: aleksander.herczek@us.edu.pl
[b]Paleontological Institute Russian Academy of Sciences, Profsoyuznaya Street 123, 117997
Moscow, Russia; e-mail: popovpin@gmail.com, lab@palaeoentomolog.ru
[c] Schmalhausen Institute of Zoology the National Academy of Sciences of Ukraine
15 Bodan Khmelnitsky Street, Kiev 01-601, Ukraine; e-mail: perkovsk@gmail.com

---

Abstract
The second representative of a new isometopine species *Archemyiomma schaeferi* sp. n. is described from the Eocene Ukrainian (Rovno) amber. The finding of a species from the same genus *Archemyiomma* Herczek, 1993 is an additional proof that Rovno (Ukraine) and Baltic ambers are of Eocene age. The inclusion belongs to the collection of the Schmalhausen Institute of Zoology of the National Academy of Sciences of Ukraine in Kiev.

Keywords
Heteroptera; Miridae; Isometopinae; *Archemyiomma* sp. n.; Rovno (Ukraine) amber

Introduction

Rovno amber is found predominantly in the lowest Oligocene, but similarity of dominant species with Baltic amber found suggests a late Eocene age (Perkovsky et al. 2010). The specific composition of the fossil fauna preserved in Rovno amber (more than 100 species not found in Baltic amber) indicates that the origin of this kind of amber was independent from that of Baltic amber (Perkovsky et al. 2003; Dlussky & Rasnitsyn 2009). However, some scientists believed that the Ukrainian amber was part of the re-deposited Baltic amber (Katinas 1971, 1987). Nine heteropteran families are found in Rovno amber. They are: Anthocoridae, Lygaeidae, Microphysidae, Miridae, Piesmatidae, Saldidae, Schizopteridae, Reduviidae and Tingidae (Perkovski et al. 2010; Popov et al. 2011). Most probably Rovno amber is of the same age as Baltic amber (Kosmowska-Ceranowicz 1999; Perkovsky et al. 2003, 2010).

Some authors, Dlussky & Rasnitsyn (2009), for instance, demonstrated that the ants known from Baltic, Bitterfeld, Scandinavian and Rovno ambers shared 17 species (more than 80% of the total number of ant inclusions). An additional 15 species of ants described from Rovno amber were found in other listed deposits. We can also note the presence of certain common heteropteran species both in Baltic and Rovno amber, like *Electrocoris brunneus* Usinger, 1942 (Cimicoidea) or *Heissiana serafini* Popov, 2001 (Piesmatidae). In addition, there are common genera in Baltic and Ukrainian ambers as well, e.g., *Jordanofulvius* Carvalho, 1954 and *Archemyiomma* (Miridae), *Loricula* Curtis, 1833, (Microphysidae), or *Lyctoferus* Popov, 2003 (Anthocoridae), and some other (Popov, pers. commun.) Thus, the similarity of inclusions indicates that the age of ambers is also similar. On the other hand, comparison of the representative amber samples from the Baltic (Yantarny quarries) and Rovno (Klesov and Dubrovitsa) suggests that the origin of their faunas was independent (Perkovsky et al. 2007). For instance, scarcity of aphids (Hemiptera) in the Rovno amber implies a warmer, possibly subtropical climate of the Ukraine Shield (Cristalline Rock Massif) in the Late Eocene, while a greater share of Scelionidae (Hymenoptera) in the Rovno assemblage points to a more xeric environment there in comparison with the Baltic amber where Diapriidae (Hymenoptera) are more abundant (Perkovsky et al. 2007). The relative rarity of Diapriidae (Hymenoptera) corresponds to the relative abundance of some xerophylic heteropteran groups, e.g., some Reduviidae (Emesinae or Centrocneminae), Piesmatidae and some other bugs which are also typical for dry subtropical, e.g., Tingidae (Cantacaderinae, Tinginae), and wet tropical zones, e.g., Miridae (most Cylapinae, Psallopinae, some Isometopinae and some others). Unfortunately, scarcity of heteropteran inclusions, especially from Rovno amber, does not yet allow us to make final conclusions about true relationships between amber from these areas, i.e., whether they are of the same origin or whether they come from different parts of the same basin.

Taxonomy

Order Hemiptera Linnaeus, 1758
Suborder Heteroptera Latreille, 1810
Infraorder Cimicomorpha Leston, Pendergrast & Southwood, 1954
Family Miridae Hahn, 1833
Subfamily Isometopinae Fieber, 1860
Archemyiomma Herczek, 1993

Type species

Archemyiomma carvalhoi Herczek, 1993
Archemyiomma: Herczek & Popov, 2012: 64

Diagnosis

Body length about 3 mm; dorsal surface somewhat rough, usually in deeper and denser microstructure, and covered with dense, suberect hairs; general coloration dark brown to pale brown; head not less than twice as wide as long; frontal-clypeal part elongate, not developed laterally; clypeus clearly differentiated; large eyes weakly prominent; ocelli large, touching inner sides of eyes at basal part of head (at base of vertex); vertex strongly narrow, almost 2x narrower than dorsal width of eye; antennal bases set on lower level of eyes, segment II longest, IV distinctly longer than III, combined length of segments III and IV clearly longer than segment II; pronotal collar rather narrow, calli not defined, intercallian pit absent, posterior margin slightly convex; mesoscutum narrowly exposed, significantly shorter than scutellum, lateral margins pale; hemelytra roughly puctated, claval commissure quite long, longer than mesoscutum and scutellum taken together; tarsi 2-segmented.

Composition

Two species: *A.carvalhoi* from Baltic and *A. schaeferi* sp.n. from Rovno (Ukraina) ambers.
Archemyiomma schaeferi Herczek & Popov sp. n. (Figs 1–4)

Description (holotype)

Female. General coloration dark-brown, I and II antennal segments, legs and also mesoscutum gold-yellow, scutellum slightly darker, cuneus pale, most cuneal apex dark, femora and tibiae of all pairs partly dark., rostrum dark. Dorsal surface of body deeply punctuate (except mesoscutum and scutellum) and covered with dense, suberect hairs directed backward. Head from above 2.1 times as wide as long; eyes yellowish-golden; trapezoid pronotum 1.75 times as wide as long, lateral margins weakly sinuate (emarginated), mesoscutum very short, 4.5 times shorter than scutellum; claval commissure 1.3 times longer than mesoscutum and scutellum taken together. Hemelytral proportion of hemelytron, exocorium and cuneus 2.32: 1.45: 0.34.

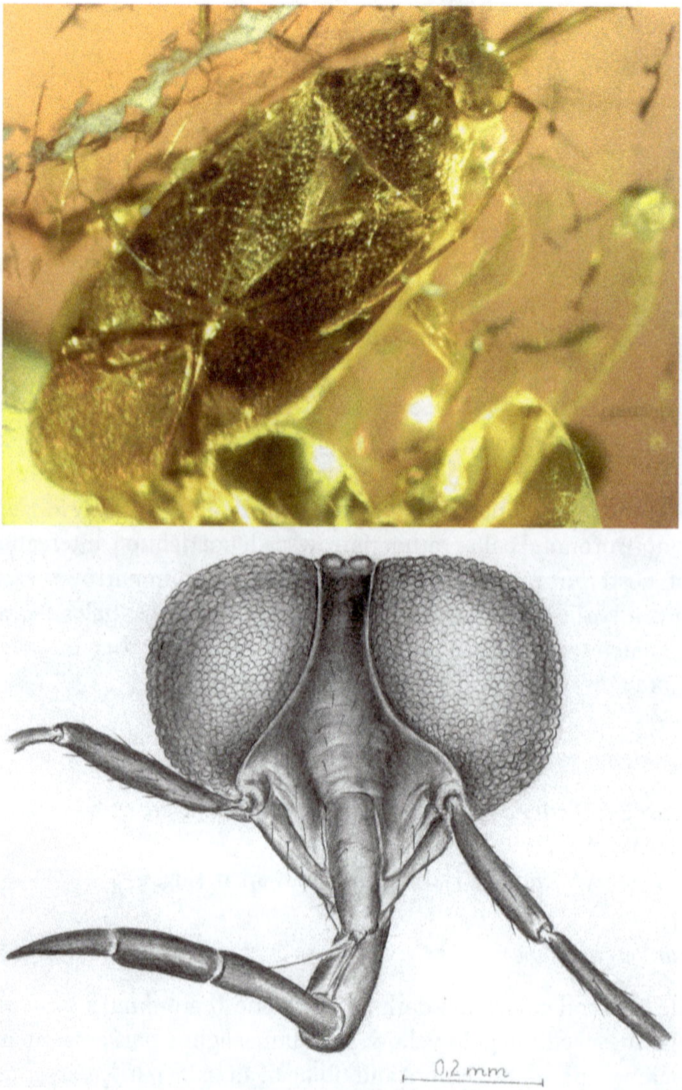

Figures 1–3. *Archemyiomma schaeferi* sp.n. (Miridae, Isometopinae), holotype, Nr. K-24175: (1) body, dorsal view; (2) idem, head, frontal view; (3) idem, body, dorsal view.

Measurements (type, all in mm)

Male: body length 3.04, width 1.06; length of head 0.27, width 0.57, height 0.62; with of vertex 0.13; width of eye 0.25, height 0.36; length of antennal segments 1–4 equal to 0.26: 0.72: 0.31: 0.52; rostral segments: equal to

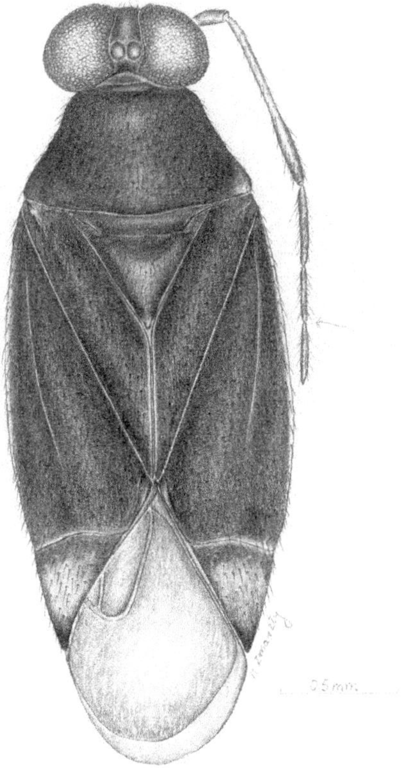

Figures 1–3. (*Cont.*)

0.25: 0.28: 0.28: 0.28; pronotum: maximum length 0.49, minimum length 0.39, anterior width 0.47, posterior width 0.86; mesoscutum length 0.08, width 0.73; scutellum length 0.36, width 0.44; claval commissure 0.60; hind leg: length of femur 0.81, width 0.2, tibia length 1.2, width 0.04; tarsus 0.33 (0.09: 0.25).

Etymology

Named after our friend and colleague Prof. Carl Schaefer (Storrs, CT, USA), eminent heteropterologist who made a great contribution to the study of the suborder Heteroptera.

Material examined

Macropterous female. Holotype Nr. K-24175 Klesov, Rovno amber; a well preserved and complete bug is included in an amber elongate almost

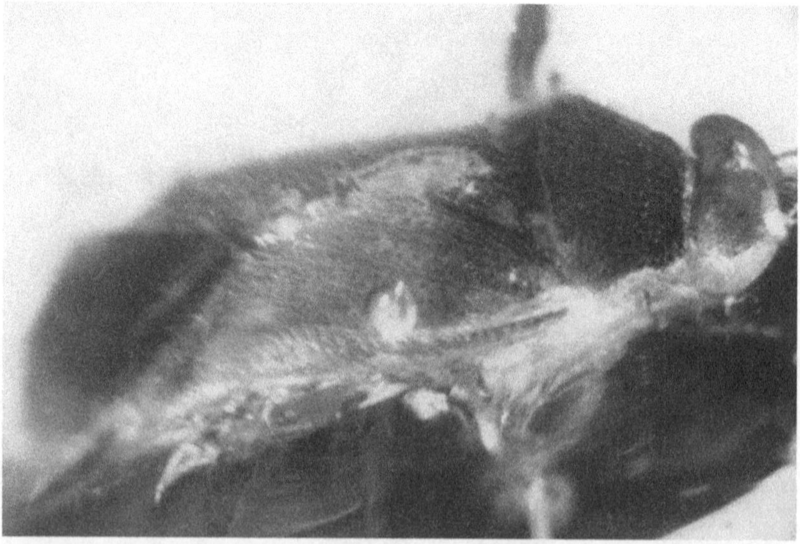

Figure 4. *Archemyiomma carvalhoi* holotype, In. 181000, Samland, No. 478, Natural History Museum, London.

parallelepiped (4.0×16.0×7.0 mm) with a piece of dark organic matter and gas bubbles; the ventral surface is covered by a 'milky' layer; housed at the Schmalhausen Institute of Zoology, National Academy of Sciences of Ukraine, Kiev.

Discussion

The first record of fossil Isometopinae was published by Hurd, Smith and Durcham (1962) from the Miocene Mexican amber and the same was cited later on by Kielbach (1982) and Spahr (1988). The genus *Archemyiomma* was one of the first fossil genera of the subfamily Isometopinae described from the Eocene Baltic amber (Herczek 1993). This new species like *Archemyiomma carvalhoi* (we also know another specimen of *A. carvalhoi* from the collection of E. Heiss, HE 07-03) also possesses some main generic characters: weakly prominent eyes, IV antennal segment distinctly longer than III, very narrow vertex, quite large ocelli, pronotum without distinct calli and slightly convex posterior margin of pronotum, longer claval commissure than combine length of mesoscutum and scutellum, and also distinct rough dorsal surface of body covering with dense hairs nonetheless, *A. scheferi* is easily distinguished by the following features: general coloration dark-brown (type species has black head, pronotum, hemelytral membrane and also dark apex of cuneus), mesoscutum and scutellum polished (TS with scutellum deeply punctate), rostral segments II–IV are equal size (TS with second rostral segment longest), pronotum about 1.75 times as broad as long (TS, ? twice as broad as long), mesoscutum 4.5 times shorter than scutellum (TS with mesoscutum 3 times shorter than scutellum), cuneus 7.5 times shorter than corium (TS has cuneus shorter 5 times than corium). The species is also some slender (length of body 2.9 times as long as broad) than type species of *A. carvalhoi* (body 2.2–2.35 times as long as broad).

In addition to the differences between these close species mentioned above some climatic and ecological distinctions of Ukranian (Rovno) and Baltic ambers may also have contributed to existence of these two different species. There is another representative of this genus from Bitterfeld amber which most probably belongs to type species *A. carvalhoi* (e.g., body twice as long as broad and pronotum twice as broad as length), but segment II of the rostrum unusually the shortest.

Acknowledgements

The authors are thankful to Prof. Dany Azar for his kind invitation to contribute to this special book and to reviewers who made some linguistic improvements. We thank Marzena Zmarzły (Silesian University, Department of Zoology, Katowice, Poland) for the excellent drawings of figures.

References

Dlussky, G.M. & Rasnitsyn, A.P. (2009) Ants (Insecta: Vespida: Formicidae) in the Upper Eocene amber of Central and Eastern Europe. *Paleontolological Journal* **43**: 1024–1042.

Herczek, A. (1993) Systematic position of Isometopinae (Miridae, Heteroptera) and their intrarelationships. *Wydawnictwo Uniwersytetu Śląskiego, Katowice* **1357**: 1–88.

Hurd, P.D., Smith, R.F. & Durham, J.W. (1962) The fossiliferous amber of Chiapas, Mexico. *Ciencia* **21**: 107–118.

Katinas, V.I. (1971) *Amber and amber-bearing sediments in Southern Baltic area.* Mintis, Vilnius, 156 pp. (in Russian).

Katinas, V.I. (1987) In: *The amber-bearing terrigenous-glauconitic formation of the Paleogene of Baltic area and Belorussia. Tectonics, facies and formations of the western part of East European Platform.* Nauka i tekhnika, Minsk, pp. 184–189. (in Russian).

Kosmowska-Ceranowicz, B. (1999) Succinite and some other fossil resines in Poland and Europe (deposits, finds, features and difference in IRS). *Esudios del Museol de Ciencias Naturales de Alava* **14**: 73–117.

Perkovsky, E.E., Rasnitsyn, A.P., Vlaskin, A.P. & Taraschuk, M.V. (2007) A comparative analysis of the Baltic and Rovno amber arthropod faunas: representative samples. *African Invertebrates* 48: 229–245.

Perkovsky, E.E., Zosimovich, V.Y. & Vlaskin, A.Y. (2003) Rovno amber fauna: a preliminary report. *Acta zoological Cracovensia*, Kraków **46** (Suppl. Fossil Insects): 423–430.

Perkovsky, E.E., Zosimovich, V.Y. & Vlaskin, A.Y. (2010) Rovno amber. In: Penney, D. (Ed.) *Biodiversity of fossils in amber from the major world deposits*. Syri Scientific Press, Manchester, pp. 116–136.

Popov, Yu.A., Kosmowska-Ceranowicz, B., Herczek, A. & Kupryjanowicz, J. (2011) Review of true bugs (Insecta: Hemiptera, Heteroptera) from the amber collection of the Museum of the Earth of PAS In Warsaw with some remarks on heteroptera insects from Eocene European amber. *Polish Journal Entomology* **80**: 699–728.

Spahr, U. (1988) Ergänzungen und Berichtigungen zu R.Kielbach's Bibliographie und Liste der Bernsteinfossilen. Überordnung Hemipteroidea. *Stuttgarter Beiträge zur Natutkunde, Serie B (Geologie und Paläontologie)* **144**: 1–60.

On the systematic position of the genera *Lepiceroides* gen.n. and *Haplochelus*, with notes on the taxonomy and phylogeny of the Myxophaga (Coleoptera)

Alexander G. Kirejtshuk[a,b] **and George Poinar Jr**[c]

[a]Zoological Institute of the Russian Academy of Sciences, Universitetskaya emb. 1, St. Petersburg, 199034, Russia; e-mails: kirejtshuk@gmail.com, agk@zin.ru
[b]CNRS UMR 7205, Muséum National d'Histoire Naturelle, CP 50, Entomologie, 45 rue Buffon, F-75005 Paris, France
[c]Department of Zoology, Oregon State University, Corvallis, OR 97331, USA
e-mail: poinarg@science.oregonstate.edu

Abstract

Lepiceroides pretiosus gen. et sp. n. (Coleoptera: Lepiceridae), which is closely related to the recent species of *Lepicerus* Motschusky, 1855, is described from Lower Cretaceous Burmese amber. The two genera differ from each other in the shape of the labrum, maxillary palpi, number of antennomeres, elytral sculpture, separation of the meso- and metacoxae and shape of the tibiae. The distinctiveness of the genus *Haplochelus* within the superfamily Lepiceroidea (Myxophaga) and its unique characteristics as related to the order Coleoptera are reiterated. The long and narrow gula, completely fused elytra without the trace of a suture, faint longitudinal costae, 7-segmented antennae, narrowly separated pro-, meso- and metacoxae, long mesoventrite and very short metaventrite justify the retention of the taxon in the family group with the rank of subfamily. Explanations for the wide family diversification of this group are presented. Phylogenetic scenarios regarding the age and origin of the Myxophaga are proposed.

Keywords

Myxophaga; Lepiceridae; new genus; new species; phylogeny; fossil record; Lower Cretaceous; Burmese amber

Introduction

The Myxophaga is a small isolated group proposed comparatively recently (Crowson, 1955) and over the past decades it was a choice topic for systematic and morphological studies. The fossil members of this group were described in the last decade from the Lower Cretaceous Burmese amber (Kirejtshuk & Poinar 2006; Kirejtshuk 2009) and from Liaoning

(Cai et al. 2012). Very recently some extinct families were included in this group (Bouchard et al. 2011) and some hypotheses were proposed regarding its "phylogeny" (Beutel 1999, 2005; Ge et al. 2010, among others). Nevertheless the fossil record of this group is still too insufficient to make reliable phylogenetic reconstructions. In this paper a new genus closely related to the recent *Lepicerus* Motschusky, 1855 is described and some comments on the evolutionary and systematic significance of the Myxophaga are presented.

Materials and Methods

The Burmese amber piece was obtained from a mine first excavated in 2001, at Hukawng Valley, southwest of Maingkhwan in Kachin State (26°20′ N, 96°36′ E) in Burma (Myanmar). This new amber site, known as the Noije Bum 2001 Summit Site, was assigned to the Early Cretaceous, Upper Albian, on the basis of paleontological evidence (Cruckshank & Ko 2003), placing the age at 97 to 110 mya. Nuclear magnetic resonance (NMR) spectra and the presence of araucaroid wood fibers in amber samples from the Noije Bum 2001 Summit site indicate an araucarian (possibly *Agathis*) tree source for the amber (Poinar et al. 2007). For study ordinary optic equipment was used, in particular the stereomicroscope Leica MZ 16.0 in the Zoological Institute of the Russian Academy of Sciences, St. Petersburg; stereomicroscope Olympus SCX9 and inverted microscope Olympus CK 40 in the Muséum National d'Histoire Naturelle Museum; and also stereomicroscope Nikon SMZ-10 in the Oregon State University, Corvallis.

Description

Family Lepiceridae Hinton, 1936
Subfamily Lepicerinae Hinton, 1936
Genus *Lepiceroides* gen. nov.

Type species

Lepiceroides pretiosus sp. n.

Etymology

The name of the new genus is formed from the generic name "*Lepicerus*" and the Greek "*oides*" ("οιδες", like resembling similar to). The gender is masculine.

Diagnosis

The new genus differs from other members of the subfamily (*Lepicerus*) in possessing a very large subquadrangular labrum, rather short maxillary palpi, 5-segmented antennae, narrower prosternal process, four rows of tubercles (not 2–3 carinas) on each elytron, elytral sides without excavations for intermediate legs (but with deep excavations for posterior legs), metacoxae more narrowly separated than mesocoxae and tibiae at least twice as wide as tarsi.

Description

Body small, under 2 mm in length, rather convex dorsally and subflattened to concave ventrally; dorsum with very coarse and slightly irregular sculpture, represented on elytra by distinct tubercles arranged in five rows, without visible pubescence; frons ventrally oriented, rather reduced mouthparts dislocated ventrally and concealed by very large subquadrate labrum with median projection; maxillary palpi very short; antennae five-segmented, with ultimate antennomere subtriangular and widest at subtruncate apex; pronotum with coarse tubercles formed along middle four pairs of paramedian ones; thorax with rather wide prosternal process, moderately long mesoventrite and rather short metaventrite; all coxae moderately large: procoxae rather narrowly separated; mesocoxae more widely separated than metacoxae; legs moderately raised, with distinct trochanter and spinose tibiae, tarsi one-segmented, long and straight, with large terminal, scarcely curved single claw; abdomen with six apparent ventrites; elytra with four rows of coarse tubercles completely covering abdomen; epipleura with deep excisions for posterior legs.

Lepiceroides pretiosus sp. n. (Figs 1–7)

Etymology

The epithet of the new species means “valuable”, “important”, “precious”.

Material

Holotype in Burmese amber, deposited in the Poinar amber collection (Accesion No. B-C-46) maintained at Oregon State University. The specimen is in a small elongate and irregular piece of amber together with many very small gas bubbles and fragments of dark organic matter. It was

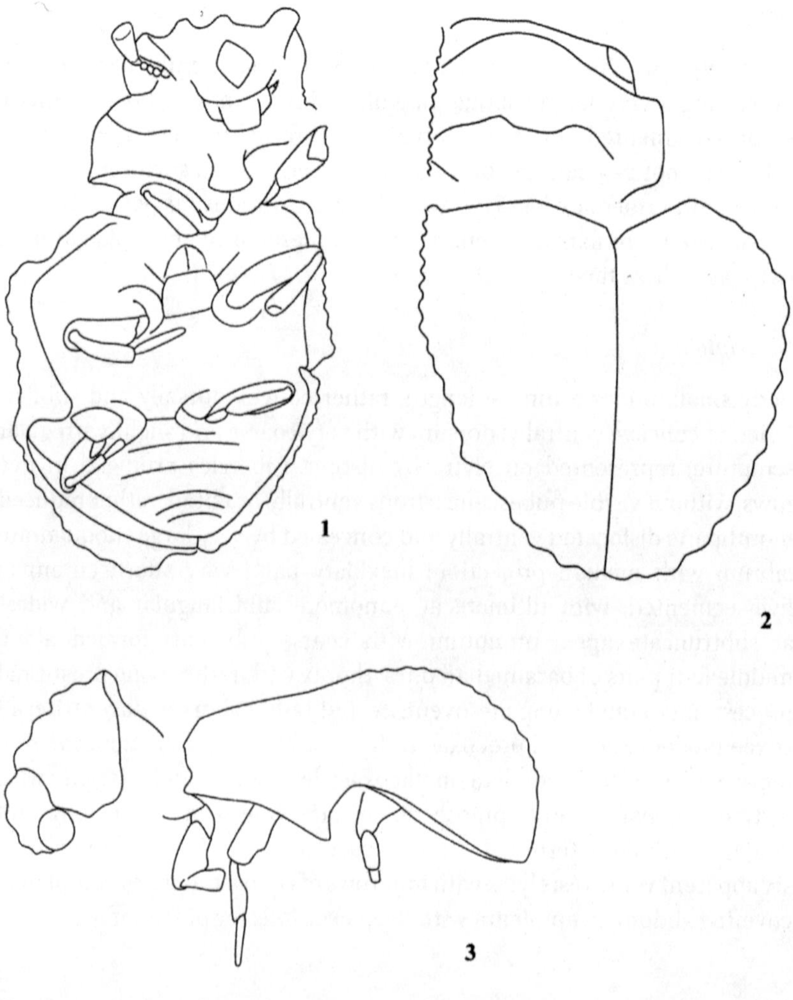

Figures 1–3. *Lepiceroides pretiosus* gen. et sp. n., holotype, body. (1) Ventral view, (2) dorsal view, (3) lateral view. Length of specimen 1.4 mm.

subsequently placed in Canada balsam and mounted between two square microscope slides. It is difficult to study because of its position in the amber and its dark and somewhat deformed body.

Type locality

Lignitic seams in sandstone-limestone deposits in the Hukawng Valley, Myanmar (Burma).

4

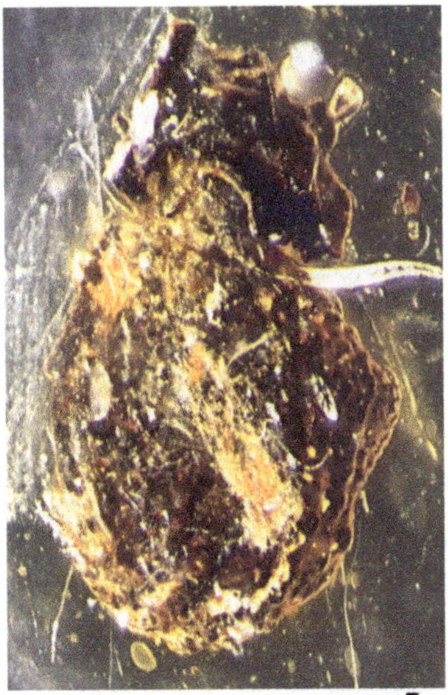

5

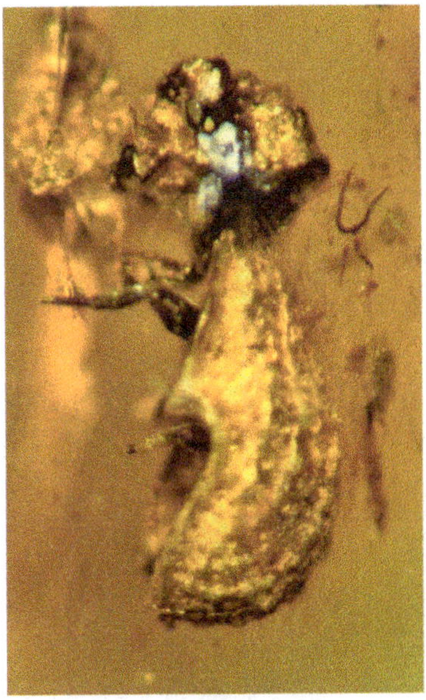

6

7

Figures 4–7. *Lepiceroides pretiosus* gen. et sp. n., holotype, body. (4–5) Dorsal view, (6) lateral view, (7) ventral view. Figures 4 and 6 made on stereomicroscope Nikon SMZ-10; figures 5 and 7 made on stereomicroscope Leica MZ 16.0. Length of specimen 1.4 mm.

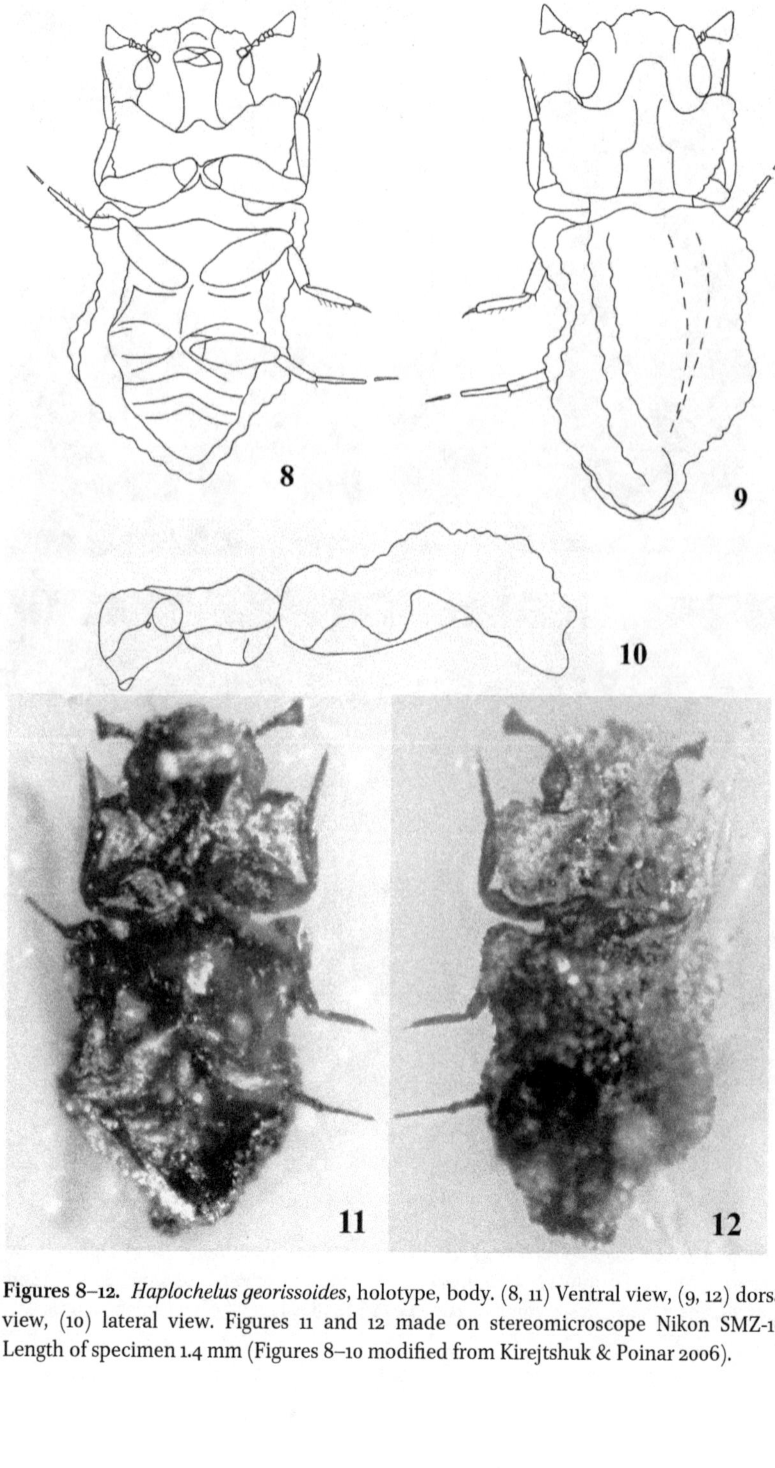

Figures 8–12. *Haplochelus georissoides*, holotype, body. (8, 11) Ventral view, (9, 12) dorsal view, (10) lateral view. Figures 11 and 12 made on stereomicroscope Nikon SMZ-10. Length of specimen 1.4 mm (Figures 8–10 modified from Kirejtshuk & Poinar 2006).

Description

Length 1.4, width 0.7, height nearly 0.3 mm (with greatest height at distal third of elytra); elongate-oval, moderately convex dorsally and subflattened to concave ventrally; subunicolorous dark brown with lighter (partly nearly yellowish) appendages; dull; integument without trace of punctations; dorsum with extremely coarse sculpture but mostly without pubescence; underside with sparse and very short, recumbent and nearly inconspicuous hairs.

Head transverse, turned ventrally with very short underside behind mouth orifice; mouthparts directed ventrad and concealed by large subquadrate labrum (about third as wide as head) with median projection; only short left penultimate and ultimate maxillary palpomeres exposed from under labrum; antennae with five antennomeres, four of them suboval and approximately of equal length, antennomere 5 (club) nearly as long as previous antennomeres combined.

Pronotum strongly transverse, apparently subparallel-sided, rather convex medially with four pairs of paramedian tubercles and very widely excavated sides; both anterior and posterior angles widely rounded; anterior edge bisinuate. Prosternum very short, with very short, wide subtruncate process at apex. Mesoventrite with a median plate in distal half and a deep concavity divided by a sharp median carina in anterior half. Procoxae not visible. Scutellum not visible. Elytra convex, each with four longitudinal rows of tubercles; not explanate at sides. Epipleura appearing as stripe with deep excavations for posterior legs. Abdomen apparently with six ventrites; ventrites 1–3 and hypopygidium comparable in length; ventrites 4 and 5 much shorter; hypopygidium widely rounded at apex.

Tibiae rather wide (somewhat narrower than femora) and slightly dorsoventrally compressed and sub-parallel sided, apparently widened at base along inner edge. Femora rather robust and apparently of usual shape (probably thickest and widest at the middle), although their apices apparently rather wide. Single tarsomere of each leg very long and subcylindrical, gradually narrowing apically, unique claw of each leg very slightly curved (approximately one third as long as tarsomere).

Discussion

Systematic placement and phylogenetic position were discussed by many coleopterists, and recently Beutel (1999) made his phylogenetic analysis for the recent representatives of the "suborder" Myxophaga.

Kirejtshuk & Poinar (2006) proposed a new family Haplochelidae for for the Cretaceous *Haplochelus*. Ge et al. (2010) attempted to "re-evaluate" the phylogenetic attempt of Beutel (loc. cit.), and concluded that *Haplochelus* should be transferred to the family Lepiceridae. However certain significant features of this genus, which separate *Haplochelus* from its closest relative (*Lepicerus* Motschusky, 1855), were not taken into consideration by Ge et al., since they were omitted from their matrix. Other characters were erroneously coded, i.e., characters 21 ("moderately broad" prosternal process for (sub) contiguous procoxae of *Haplochelus*, which are not covered or separated by a process at all), 27 ("broadly separated" for (sub) contiguous mesocoxae), or coded without evidence, i.e., 33 (metacoxal plates in *Haplochelus*). As a result, the list of characters used by Ge et al. (2010) did not include any features of *Haplochelus* that differ from other groups of Myxophaga. Therewith the characters 21 do not take into consideration that the procoxae of *Lepicerus* are covered by prosternal process as in *Lepiceroides pretious* gen. et sp. nov. Using this matrix, the authors concluded that the distinctness between these genera is not great enough to regard them as separate families. Their decision is not "objective" in the cladistic sense.

Both members of the Myxophaga described from Burmese amber previously, *Haplochelus georissoides* Kirejtshuk & Poinar, 2006 and *Burmasporum rossi* Kirejtshuk, 2009 (the latter not included in the matrix or considered by Ge et al. 2010), and also *Lepiceroides pretiosus* sp. et gen. n., have a characteristic small body size (0.8–1.4 mm long) and rather unique features that separate them from their probable recent relatives, although *Lepiceroides* gen. n. is less distinct from its recent close relatives. *Haplochelus* Kirejtshuk & Poinar, 2006 has completely fused elytra without the trace of a suture and *Burmasporum* Kirejtshuk, 2009, in contrast to other members of the "suborder" Myxophaga, possesses the ability for the prothoracic segment and head to move downwards. These apomorphic characters have different interpretations for taxonomic and phylogenetic analyses. Transformation of the thorax, which makes it possible to roll the prothoracic segment, occurs not infrequently in different polyphagan lineages. This feature is certainly not as significant as the complete fusion of the elytra, which causes crucial transformations of the thoracic sclerites, changing the type of movement and probably the physiology of *Haplochelus*. While *Burmasporum* was placed in a family with its closest relative (*Sphaerius* Waltl, 1838), a separate family was erected for *Haplochelus*. However, the description of *Lepiceroides* gen. n. that is close to the recent *Lepicerus* calls for a revision of the rank of the latter taxon.

Haplochelus (Figs 8–12) differs from the recent lepicerids in the following characters: long and narrow gula; completely fused elytra (strongly excavate medially in the anterior 2/3 and without the trace of a suture); faint longitudinal costae; seven-segmented antennae; narrowly separated to (sub) contiguous coxae in each pair; very long mesoventrite and very short metaventrite (metaventrite shorter than mesoventrite). Also, the fused elytra are strongly transformed dorsally in the apterous *Haplochelus* (concave in the anterior part) and along the epipleura (very deep excisions for movement of intermediate and posterior legs), while the known recent representatives of *Lepicerus* have moderately developed posterior wings and normal shaped elytra, although with shallow depressions on the epipleura for reception of the femora. The deeply excised elytral sides in *Haplochelus georissoides* indicate great restrictions in movement of the intermediate and posterior legs, the coxae of which approach each other. The rather unique character of this species is the very sharp, polished median ridge in the apical forth of the elytra. Besides, the very deeply excavate gula clearly visible in the holotype of *H. georissoides* could be a sequence retraction during fossilization causing a certain deformation of some body sclerites. However, the remaining characters are suitable for diagnostics. All these peculiarities provide sound evidence for the erection of a separate family. In addition, these features suggest that the Cretaceous *Haplochelus* is certainly much more specialized than the recent *Lepicerus*, which contains more plesiomorphic characters than the extinct genus. It would appear that *Lepicerus* is closer to the common ancestor than to the Cretaceous genus. In conclusion, differences between the *Lepicerus-Haplochelus* generic lineages are quite significant, which is why they could be treated in separate, but closely related families.

The structural peculiarities of the united *Lepicerus-Lepiceroides-Haplochelus* lineage are also significant (one-segmented tarsi with a single claw; head with vertical frons; mouthparts dislocated ventrally; ultimate antennomere much larger than preceding ones, subtriangularly enlarged and abrupt apically; pronotum widened anteriorly; dorsum coarsely sculptured and without distinct punctation, etc.). The synopsis of these three genera makes them rather distinct among the known Myxophaga. On the cladogram presented by Ge et al. (2010), the uniqueness of this lineage in the configuration of the Myxophaga is rather clear and supported by many characters. This cladogram agrees with the synopsis specified in the description of *Haplochelus* (Kirejtshuk & Poinar 2006) and supports the separation of the "suborder" into two superfamilies.

The difficulties encountered when comparing extant groups with extinct ones are well known. Many diagnostic characters that are stable in a particular group are frequently not accessible in fossils. As a result researchers need to make a predictable interpretation regarding the presence or absence of established characters in the fossil and decide if they are similar to those of its recent relatives. Not fulfilling this action can result in conclusions without rational explanations.

The time of the origin and main diversification of the Myxophaga is still very unclear. The only three fossil myxophagan species were described from the Lower Cretaceous of Liaoning (Northeastern China, Yixian Formation) and Burmese amber (see above). Some other compression fossils of different ages may belong to this group, but they remain undescribed because of difficulties in interpreting their characters. Even in the case of *Hydroscapha jeholensis* Cai, Short & Huang, 2012, many characters are not discernible enough for a clear interpretation to be made. Nevertheless the forms described from Burmese amber scarcely can give rise to doubt. If we recognize the age of Burmese amber as 100–105 Myr (Cruickshank & Ko 2003) and the lower boundary of the Cretaceous as 135 Myr, we might surmise that 30 Myr would be long enough for this group to diverge. On the other hand, the Myxophaga certainly should have some ancient ancestors. The long dialogues on whether Myxophaga aligns with Adephaga or Polyphaga based on structural similarities are well known. They began with Crowson (1955) and continue to the present based mostly on morphological but recently also on molecular comparisons. Analyses of different nucleotides (mitochondrial or ribosomal) led even to a greater antipathy; not only Adephaga (Caterino et al. 2002, among others) and Polyphaga (Hughes et al. 2005, among others), but also Archostemata (Hunt et al. 2007, among others) are mentioned as the closest clades to Myxophaga. Some thoughts on this subject were expresssed by Beutel (including Beutel 1999) and also in the description of *Haplochelus* (Kirejtshuk & Poinar 2006). Ponomarenko (2001) considered that the recent groups of Myxophaga could have arisen from within the schizophoroid lineage of Archostemata sensu Ponomarenko 1969. This is a general idea which needs to be documented by fossils. Beutel (2005) concluded that the Mesozoic family Catiniidae can be linked with Myxophaga because they are "relatively small body-sized" groups (indeed the known beetles in this schizophoroid family are longer than 5.0 mm and often reaching 15.0 mm). The further supposition of Ge et al. (2010) on the Jurassic root of the family Lepiceridae and Triassic origin of the "suborder" is not explained in the paper and is probably based on data from comparison of nucleotide sequences. It certainly requires further verification, including direct fossil evidence.

The adults of Myxophaga exhibit a tendency toward miniaturization with a simplification of some structures. This could be due to an incomplete development of basic features or from a pedomorphic transformation, such as a reduced number of antennomeres, unilobed maxilla, reduced number of tarsomeres, shortening of the elytra, etc. This tendency seemed to affect different organs in various related groups that maintain a rather similar bionomy (inhabiting moist substrates at banks of running water, spray zone of waterfalls or hydropetric habits, with an excepted likely terrestrial habitat for *Sphaerius africanus* Endrödy-Younga, 1997). This secondary simplification results in a comparatively rapid reduction of most recently acquired structures and, as a result, traces of common relatedness would manifest rather obscurely. This could explain why such a small group in number of species with a similar bionomy as the Myxophaga demonstrates a comparatively wide family diversification. The very wide transcontinental distribution of species of the genera *Sphaerius* and *Hydroscapha* Leconte, 1874 seems to indicate that these groups were maintaining their structural unity during a rather long time, which could support the above scenario.

Another possible explanation is that the "suborder" could be rather old in origin with the recent fauna representing only the remains of an earlier biodiversity after various extinction events. However, there are some problems with this supposition. First of all, no true Myxophaga are known earlier than the age of Lioning Yixian Formation and Burmese amber. If this was originally a diverse group one would expect earlier forms to occur under a wider range of ecological conditions. However all recent members of the "suborder" with known bionomy show a striking similar mode of life and habits, even though they vary rather greatly in structure. The occurrence of most myxophagans in Burmese amber could have some ecological explanation, however, they need a detailed analysis of other components of the Burmese amber biota.

A compromise hypothesis is also possible. The Myxophaga could be a rather old group which was initially highly specialized for occupying ecological zones similar to those inhabited by recent representatives. Inhabitants of such environments are not suitable for fossilization and therefore, the fossil record of Myxophaga is greatly minimized. This ecological specialization over time resulted in structural diversification leading to a gradual miniaturization of group members. While the body size of Cretaceous Myxophaga is comparable to recent members, we propose that larger forms were more susceptible to various extinction events. Each of the above hypotheses have merit, however they all require more evidence, especially from additional fossil material.

The Myxophaga was proposed by Crowson (1955) for a few recent families. After this proposal, Forbes (1957) pointed out some serious contradictions in this interpretation, which are based mainly on the hindwing venation and folding, and which gave reasons for alternative interpretations. Ponomarenko (1973, 2002) insisted that it was impossible to separate the Myxophaga from other Archostemata, including extinct groups, except for differences in larvae of recent representatives (not available in fossils), and regarded these recent families as a superfamily Microsporoidea in the infraorder Schizophoromorpha (Ponomarenko 2002; Kirejtshuk & Ponomarenko 2012). In most current publications the Myxophaga are frequently treated sensu Crowson with four families represented in the recent fauna or also with some extinct families from Archostemata (Bouchard et al. 2011). In the latter case, the "suborder Myxophaga" seems to lose its significance in the context of phylogeny-based systematics, because at the moment, the extrapolation of the "suborder" features of "Myxophaga" in the initial sense with extinct groups similar to other Archostemata, seems premature (Kirejtshuk et al. in press). A correct interpretation of the subordinal characteristics will be possible only after a revision of the basal Palaeozoic groups of beetles.

Thus, the proposal for *Haplochelus* as a separate taxon of family group with a separate superfamily Lepiceroidea for its relatives is justified, although for equilibrium and stability, *Haplochelus* is regarded here as a separate subfamily of the Lepiceridae. Our interpretation of the "Myxophaga" is shown in the following key.

Key to families and subfamilies of "Myxophaga"

1. Tarsi one-segmented, each segment bearing one claw; mouthparts oriented ventrally, ultimate antennomere much larger than preceding ones, subtriangular and abrupt apically; pronotum widened anteriorly; dorsum coarsely sculptured and without distinct punctation......superfamily Lepiceroidea........(Lepiceridae Hinton, 1936)....2
- Tarsi three- or four-segmented, each terminal segment bearing two claws; mouthparts oriented apically; ultimate antennomere of

various shapes, but never much larger than preceding ones, never subtriangular or abrupt apically; pronotum widest at base and more or less gradually narrowing anteriorly; dorsum variable in sculpture and punctation.....................superfamily Sphaerioidea..........................3

2 (1). Elytra with distinct suture and distinct longitudinal costae; scutellum clearly exposed; gular sutures widely separated; protrochanter rather transverse and moderately developed; antennae 4–5-segmented; prosternal process rather wide, separating procoxae; meso- and metacoxae moderately separated; mesoventrite moderately short and markedly shorter than metaventrite; pronotum with distinct top of anterior and posterior angles. ....Lepicerinae sensu str.

- Elytra completely fused, without trace of suture; without longitudinal costae (except for a rather sharp ridge dislodged elytral suture at distal third); scutellum not exposed; gular sutures rather narrowly separated; protrochanter narrow and rather long; antennae seven-segmented; procoxae not separated by prosternal process; meso- and metacoxae narrowly separated to (sub) contiguous; mesoventrite longer than metaventrite; pronotum with rounded anterior and posterior angles ....Haplochelinae Kirejtshuk & Poinar, 2006, stat. nov.

3 (1). Tarsi 4–5 segmented; metacoxae without femoral plates; abdomen with 4–5 ventrites; elytra with longitudinal furrows (recent forms in Western Hemisphere) or without (recent forms in South Africa and Eastern South Asia); legs comparatively long with rather narrow subparallel- sided tibiae; antennae 9- or 11-segmented with urn-shaped structure of antennomeres 1 and 2..................Torridincolidae Steffan, 1964 4

- Tarsi three-segmented; metacoxae with femoral plates; abdomen with three or 6–7 ventrites; elytra without longitudinal furrows; legs not long with somewhat dilated tibiae; antennae 5–11-segmented but never with urn-shaped structure of antennomeres 1 and 2.........5

4 (3). Antennae nine-segmented; body contracted between pronotum and elytra; elytra with serial punctution; dorsum and underside bare; integument with extensive sculptural elements .... Torridincolinae sensu str.

- Antennae 11-segmented; body regularly oval; elytra without serial punctution; dorsum and underside finely pubescent; integument without pronounced sculptural elements .... Deleveinae Endrödy-Younga, 1997

5 (3). Abdomen with three ventrites; metacoxae oblique and contiguous; elytra completely covering abdomen; antennae 11- segmented with 3–4-segmented club; posterior wings with "oblongum" .... Sphaeriusidae Erichson, 1845

- Abdomen with six ventrites; metacoxae transversely disposed and more or less separated; elytra not complete and leaving apex of abdomen uncovered; antennae 5–8 segmented with one-segmented club; posterior wings without "oblongum"..Hydroscaphidae LeConte, 1874

Acknowledgements

The first author was supported by the Programme of the Presidium of the Russian Academy of Sciences "Problems of the Origin of Life and Formation of the Biosphere" and a grant from the Russian Foundation for Basic Research (12-04-00663-a). Rather important assistance was received from Dany Azar (Lebanese University, Sciences, Fanar–Matn, Lebanon) who helped the authors in preparation of the specimen for study. The authors appreciate comments and suggestions from several anonymous reviewers. The junior author thanks Alex E. Brown for donating the specimen of *Lepiceroides pretiosus* gen. et sp. n., to the Poinar amber collection.

References

Beutel, R.G. (1999) Phylogenetic analysis of Myxophaga (Coleoptera) with a redescription of *Lepicerus horni* (Lepiceridae). *Zoologischer Anzeiger* 237: 291–308.

Beutel, R.G. (2005) Introduction, phylogeny. In: Beutel, R.G. & Leschen, R.A.B. (Eds) *Handbook of Zoology Volume IV. Arthropoda: Insecta. Coleoptera, Volume 1: Morphology and Systematics (Archostemata, Adephaga, Myxophaga, Polyphaga partim)*. Walter De Gruyter, Berlin, pp. 43–44.

Bouchard, P., Bousquet, Y., Davies, A.E., Alonso-Zarazaga, M.A., Lawrence, J.F., Lyal, C.H.C., Newton, A.F., Reid, C.A.M., Schmitt, M., Ślipiński, S.A. & Smith, A.B.T. (2011) Family-group names in Coleoptera (Insecta). *ZooKey* 88: 1–972.

Cai, C., Short, E.Z. & Huang, D. (2012) The first skiff beetle (Coleoptera: Myxophaga: Hydroscaphidae) from Early Cretaceous Jehol Biota. *Journal of Paleontology*, 86: 116–119.

Caterino, M.S., Shull, V.L., Hammond, P.M. & Vogler, A.P. (2002) Basal relationships of Coleoptera inferred from 18S rDNA sequences. *Zoologica Scripta* 31: 41–49.

Crowson, R.A. (1955) *The natural classification of the families of Coleoptera*. Nathaniel Lloyd, London, 187 pp.

Cruickshank, R.D. & Ko, K. (2003) Geology of an amber locality in the Hukawng Valley, northern Myanmar. *Journal of Asian Earth Sciences* **21**: 441–455.

Forbes, W.T.M. (1957) Note on the Coleoptera Myxophaga. *The Coleopterists' Bulletin* **11**: 12.

Ge, S.-Q., Friedrich, F. & Beutel, R.G. (2010) On the systematic position and taxonomic rank of the extinct myxophagan †*Haplochelus* (Coleoptera). *Insect Systematics & Evolution* **41**: 329–338.

Hughes, J., Longhorn, S.J., Papadopoulou, A., Theodorides, K., Riva, A. de, Mejia-Chang, M., Foster, P.G., & Vogler, A.P. (2006) Dense taxonomic EST sampling and its applications for molecular systematics of the Coleoptera (beetles). *Molecular Biology and Evolution* **23**: 268–278.

Hunt, T., Bergsten, J., Levkanikova, Z., Papadopoulou, A., John, O.S., Wild, R., Hammond, P.M., Ahrens, D., Balke, M., Caterino, M.S., Gómez-Zurita, J., Ribero, J., Barracllough, F.G., Bokakova, M., Bokak, L. & Vogler, A.P. (2007) A comprehensive phylogeny of beetles reveals the evolutionary origins of a superradiation. *Science* **318**: 1913–1916.

Kirejtshuk, A.G. (2009) A new genus and species of Sphaeriusidae (Coleoptera, Myxophaga) from Lower Cretaceous Burmese amber. *Denisia* N.S. **86**: 99–102.

Kirejtshuk, A.G. & Poinar Jr., G. (2006) Haplochelidae, a new family of Cretaceous beetles (Coleoptera, Myxophaga) from Burmese amber. *Proceedings of the Entomological Society of Washington* **108**: 155–164.

Kirejtshuk, A.G., Poschmann, M., Prokop, J., Garrouste, R. & Nel, A. (in press) Evolution of the elytral venation and structural adaptations in the oldest Palaeozoic beetles (Insecta: Coleoptera: Tshekardocoleidae). *Journal of Systematic Palaeontology*.

Kirejtshuk, A.G. & Ponomarenko, A.G. (2012) *Catalogue of fossil Coleoptera*. Available online at http://www.zin.ru/Animalia/Coleoptera/rus/paleosys.htm (accessed September 2012).

Poinar Jr., G.O., Lambert, J.B. & Wu, Y. (2007) Araucarian source of fossiliferous Burmese amber: spectroscopic and anatomical evidence. *Journal of the Botanical Research Institute of Texas* **1**: 449–455.

Ponomarenko, A.G. (1969) Istoricheskoe razvitie zhestkokrylykh-arkhostemat [Historical development of the Coleoptera – Archostemata.] *Trudy Paleontologicheskogo Instituta Akademiya Nauk SSSR*, **125**: 240 pp. (in Russian)

Ponomarenko, A.G. (1973) O delenii otryada zhestkokrylykh na podotrydy [On subdividing of the order Coleoptera into suborders. In: Narchuk, E.P. (Ed.) *Voprosy paleontologii nasekomykh* [Problems of the Insect Palaeontology]. *Doklady na 24-m Ezhegodnom chtenii pamyati N.A.Kholodkovskogo, 1971* [Lectures on the XXIV Annual Readings in Memory of N.A. Kholodkovsky (1–2 April 1971)]. Leningrad, Nauka, pp. 78–89. (in Russian)

Ponomarenko, A.G. (2001) Superorder Scarabaeoidea Laicharting, 1781. Order Coleoptera Linne, 1758. The beetles. In: Rasnitzyn, A.P. and Quicke, D.L.J. (Eds) *History of insects*. Kluwer, Dordrecht, pp. 164–180.

Ponomarenko, A.G. (2002) Superorder Scarabaeidea Laicharting, 1781. Order Coleoptera Linné 1758. In: Rasnitsyn, A.P. & Quicke, L.J. (Eds): *History of Insects*. Kluwer, Dordrecht, pp. 164–176.

PART TWO

INSECTS FROM UPPER CRETACEOUS AMBER

Two new biting midges of the modern type from Santonian amber of France (Diptera: Ceratopogonidae)

Joanna Choufani[a], Vincent Perrichot[b], Vincent Girard[c,d], Romain Garrouste[a], Dany Azar[e] and André Nel[a]

[a]CNRS UMR 7205, Muséum National d'Histoire Naturelle, CP 50, Entomologie, 45 rue Buffon, F-75005 Paris, France.
E-mail: choufanijoana@hotmail.fr, garroust@mnhn.fr, anel@mnhn.fr
[b]CNRS UMR 6118 Géosciences & OSUR, Université de Rennes 1, Campus de Beaulieu bât. 15, 263 avenue du Général Leclerc, 35042 Rennes cedex, France.
E-mail: vincent.perrichot@univ-rennes1.fr
[c]Centre de Bio-Archéologie et d'Ecologie (UMR 5059 CNRS/Université Montpellier 2/ EPHE/INRAP), Institut de Botanique, 163 Rue Auguste Broussonet, 34090 Montpellier, France. E-mail: vincent.girard@univ-montp2.fr
[d]Université Montpellier 2, Place Eugène Bataillon, 34095 Montpellier, France.
[e]Lebanese University, Faculty of Sciences II, Department of Biology, P.O. Box 26110217, Fanar – Matn, Lebanon. E-mail: azar@mnhn.fr

Abstract

The second Mesozoic representative of the (Dasyheleinae & Forcipomyiinae) clade of Ceratopogonidae, *Devalquia brisaci* gen. et sp. n., as well as the first Mesozoic Ceratopogoninae Heteromyiini *Metahelea roggeroi* sp. n., are the first arthropods described from the Santonian amber outcrop of Piolenc (Vaucluse, France). Both confirm the great morphological stability through time within the Ceratopogonidae. They also suggest that the other ceratopogonine clades were also present by at least the Upper Cretaceous.

Keywords

Ceratopogonidae, Forcipomyiinae, Dasyheleinae, Ceratopogoninae, Heteromyiini, gen.n., sp.n., Santonian, amber

Introduction

Ceratopogonidae, or biting midges, are a family of small nematoceran Diptera with more than 5000 described living species. They are closely related to the Chironomidae, Simuliidae, and Thaumaleidae. Their earliest record is from the Early Cretaceous amber of Lebanon, but the high diversity reported from this fossil material suggests a much earlier origin (Borkent 2000, 2001; Szadziewski 1996). Their minute size could explain the reason of their absence or overlooking in Triassic or Jurassic fossil insect taphocenoses, as their state of preservation within the sediment would not allow their recognition. Nevertheless, occurrences in Triassic or Jurassic amber remain possible, as recently demonstrated by the discovery of an undetermined midge in the Triassic amber of Italy (Azar et al. 2010; Schmidt et al. 2012). Amber remains the best medium for the preservation of fossil ceratopogonids and other minuscule organisms, with exquisite details available for the taxonomical study, sometimes as if these inclusions entombed in resin millions of years ago were freshly collected material.

We describe herein the first fossil insects from the Santonian amber of Piolenc (Vaucluse department, Provence-Alpes-Côte-d'Azur region, Southeastern France). Other material comprises tiny beetles and wasps in droplets of amber. These fossils correspond to the earliest representative ceratopogonids of the tribe Heteromyiini and closely related to representatives of the Forcipomyiinae and Dasyheleinae. These minute flies were discovered by one of us (VP) and an amateur paleontologist (Daniel Roggero) in tiny droplets of amber. Piolenc is a fossiliferous outcrop yielding abundant meso- and macroremains of plants, including lignite and cuticle compressions of an angiosperm-dominated leaf-assemblage that formed under a warm, tropical climate with probable dry seasons (Gomez et al. 2003). Although the amber pieces are generally small to very small in size (generally less than 5 mm in diameter), some macro-inclusions can be found, including arthropods. While the Lower to mid-Cretaceous entomofaunas are relatively well documented thanks to extensively studied outcrops from Lebanon, France, Spain, and Myanmar; it is not the case for the Santonian. If lignite is known from Piolenc as early as in the 19th century (Dumas 1876), amber was not recorded in these publications (Lacroix 1910), but it is only in the last decade that collection and study of this amber were initiated (Onoratini et al. 2009), after the site was re-discovered by three amateur colleagues (Christian Devalque, Patrick Brisac and Daniel Roggero) who donated the material they collected themselves and guided us to the outcrop.

Materials and Methods

The amber fossils were carefully prepared in Canada balsam medium, following the method described by Azar et al. (2003), in order to observe as many characters as possible. Thus the 'absences' of structures are accurate, which is different of structures that are 'not visible' but maybe present. The specimens were examined with an Olympus SZ9 stereomicroscope and an Olympus inverted compound microscope, both equipped with camera lucida for line drawings and digital cameras.

Abbreviations used in description: AR, antennal ratio showing combined length of distal five segments X-XIV to that of basal preceding eight segments II-IX taken together; PR, maxillary palpal ratio showing length of palpal segment III to width of broadest part of segment; RL-P, relative length of five segments of maxillary palpus; RL-8A, relative length of eight distal antennal segments VII-XIV (Arabic numeral prefix before A shows number of distal antennal segments); RL-L, relative length of leg segments, except for coxa and trochanter; TR, tarsal ratio showing relative lengths of first and second tarsomeres.

Systematic Palaeontology

Order Diptera Linnaeus, 1758
Family Ceratopogonidae Newmann, 1834
Subfamily Forcipomyiinae Lenz, 1934 or Dasyheleinae Lenz, 1934

Genus *Devalquia* Choufani & Nel gen. n.

Type species

Devalquia brisaci sp. n., by present designation.

Diagnosis

Female with very hairy body; antennae with 13 unsculptured flagellomeres; distal five cylindrical and elongated, with last one having a rounded apex; palpus five-segmented with sensory pit at the tip of thickened third palpal segment; anapleural suture short; trochanter of fore and midlegs with a pair of thick setae; first tarsomere of fore leg only with slender setae; a short hind tibial spur present; wing membrane densely covered by macrotrichia; fringe complex not a single row of hairs; second radial cell not very long and narrow; costa not prolonged beyond R3; CR 0.65; claws gently curved with apices not bifid; empodia absent.

Etymology

Named after Christian Devalque for his great help in the discovery and study of the outcrop of Piolenc.

***Devalquia brisaci* Choufani & Nel sp. n.** (Figs 1–7)

Type specimen

Holotype IGR.PLC-1 (female), deposited in the amber collection of the Geology Department and Museum of the University Rennes 1, France.

Diagnosis

As for the genus (*vide supra*).

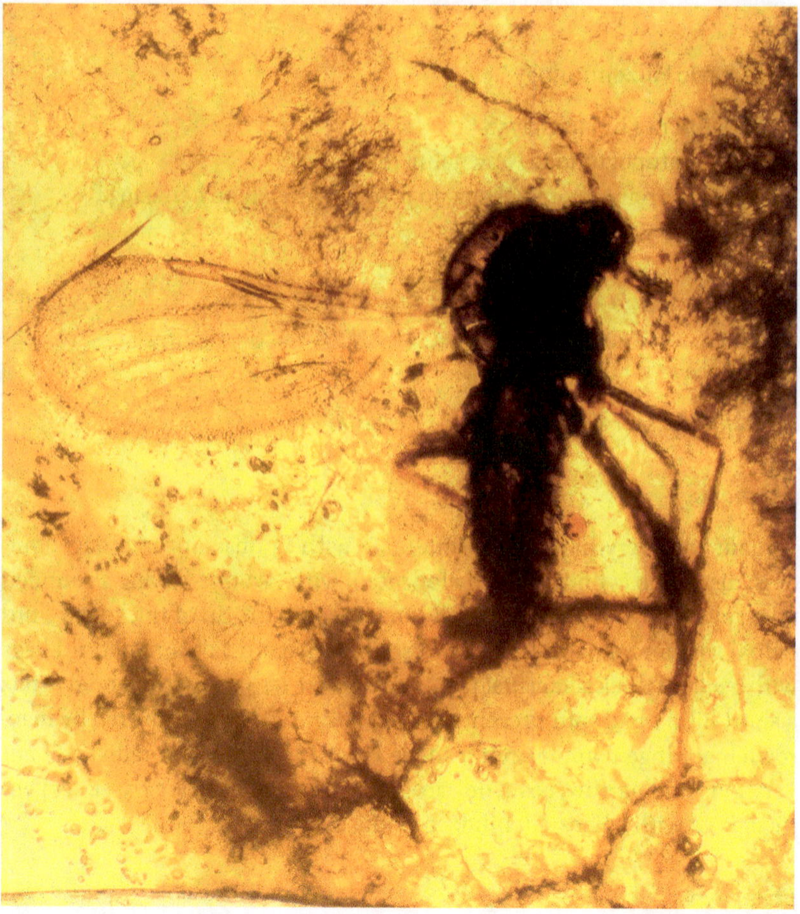

Figure 1. *Devalquia brisaci* Choufani & Nel, gen .n., sp. n., holotype specimen IGR.PLC-1, photograph of general habitus.

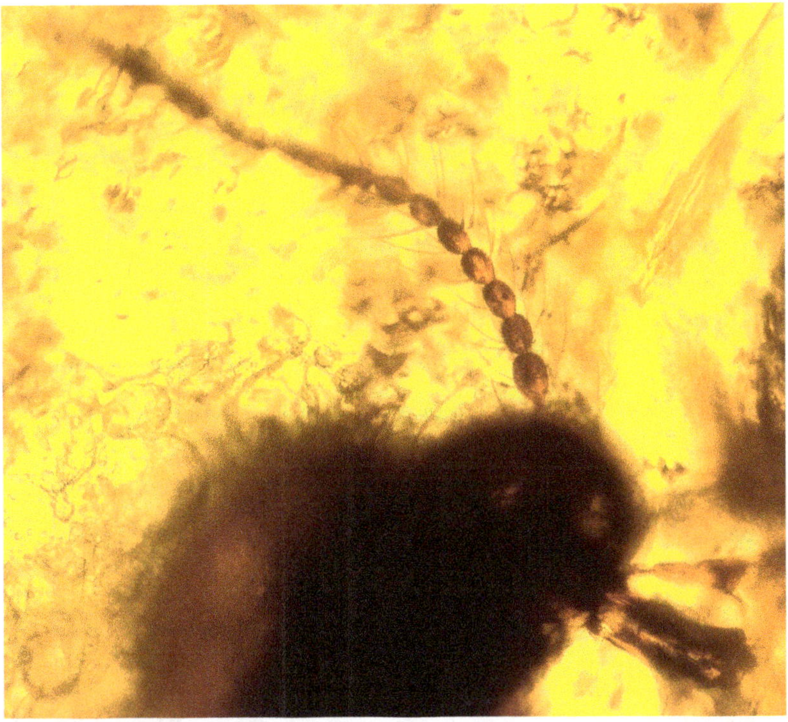

Figure 2. *Devalquia brisaci*, holotype specimen IGR.PLC-1, photograph of basal half of antenna.

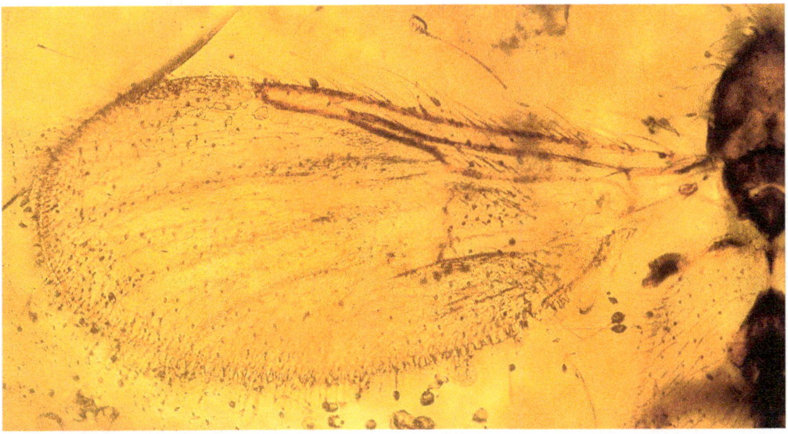

Figure 3. *Devalquia brisaci*, holotype specimen IGR.PLC-1, photograph of wing.

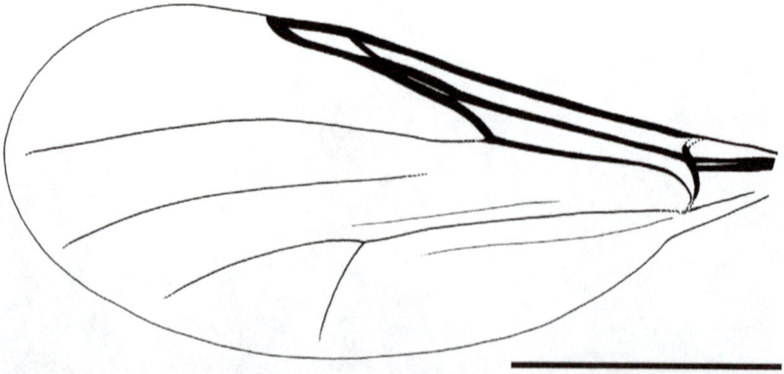

Figure 4. *Devalquia brisaci*, holotype specimen IGR.PLC-1, drawing of wing. Scale bar=0.3 mm.

Figure 5. *Devalquia brisaci*, holotype specimen IGR.PLC-1, photograph of fore tarsi with empodia and simple claws.

Age and locality

Early Santonian, sandy-to-coaly clayey beds (geological stratum 'a' in Gomez et al. 2003: fig. 1) in ancient lignite mine of Massif d'Uchaux, Piolenc, Vaucluse, France.

Etymology

Named after Patrick Brisac, who helped us to study and collect fossil in the outcrop of Piolenc.

Description

Male unknown. Body very hairy. Total body length 0.9 mm (distal part of abdomen and genitalia not visible, so it might be longer).

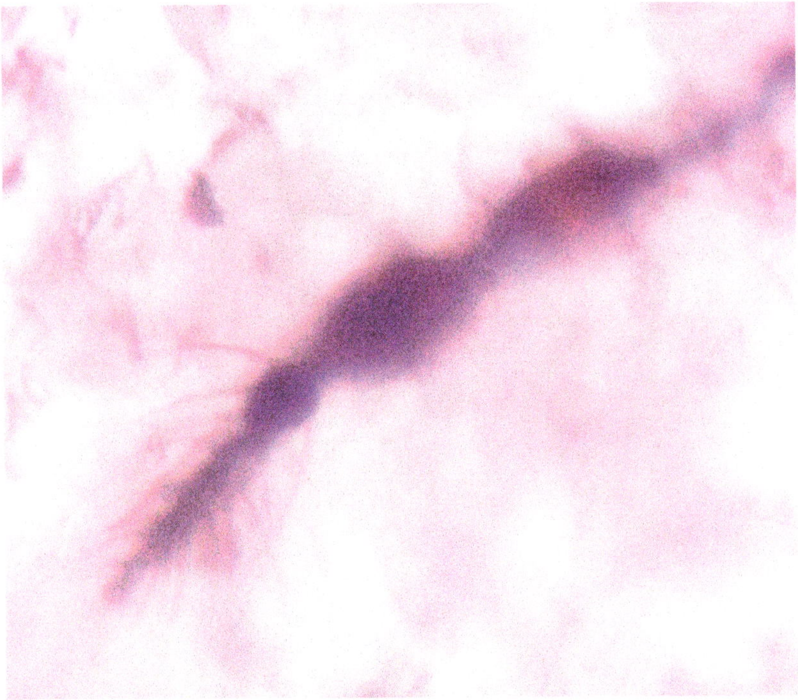

Figure 6. *Devalquia brisaci*, holotype specimen IGR.PLC-1, photograph of terminal antennomeres.

Head: bare eyes broadly contiguous; antenna with 13 unsculptured flagellomeres, proximal flagellomeres ovoid, much shorter than distal five flagellomeres cylindrical, elongated; flagellum 480 μm long; palpus five-segmented; third palpal segment thickened, 40 μm long, 26 μm wide, with sensory pit at tip; fourth and fifth segments separated, slender, equal in length 24 μm; terminal flagellomere rounded at apex.

Thorax 0.32 mm long, 0.30 mm wide, scutum and scutellum covered with very long setae, anapleural suture short.

Wing 0.875 mm long, 0.375 mm wide; first radial cell short, the second one more elongate and narrow; costa not prolonged beyond R3, CR 0.65; vein R4+5 absent; intercalary veins well visible; wing membrane densely covered with macrotrichia; fringe complex with more than a single row of hairs.

Legs slender; TR(III) 2.5; trochanter of fore and midlegs with a pair of thick setae; first tarsomere of fore leg only with slender setae; a short hind tibial spur present; claws small, gently curved, simple, equal on all legs; empodia absent.

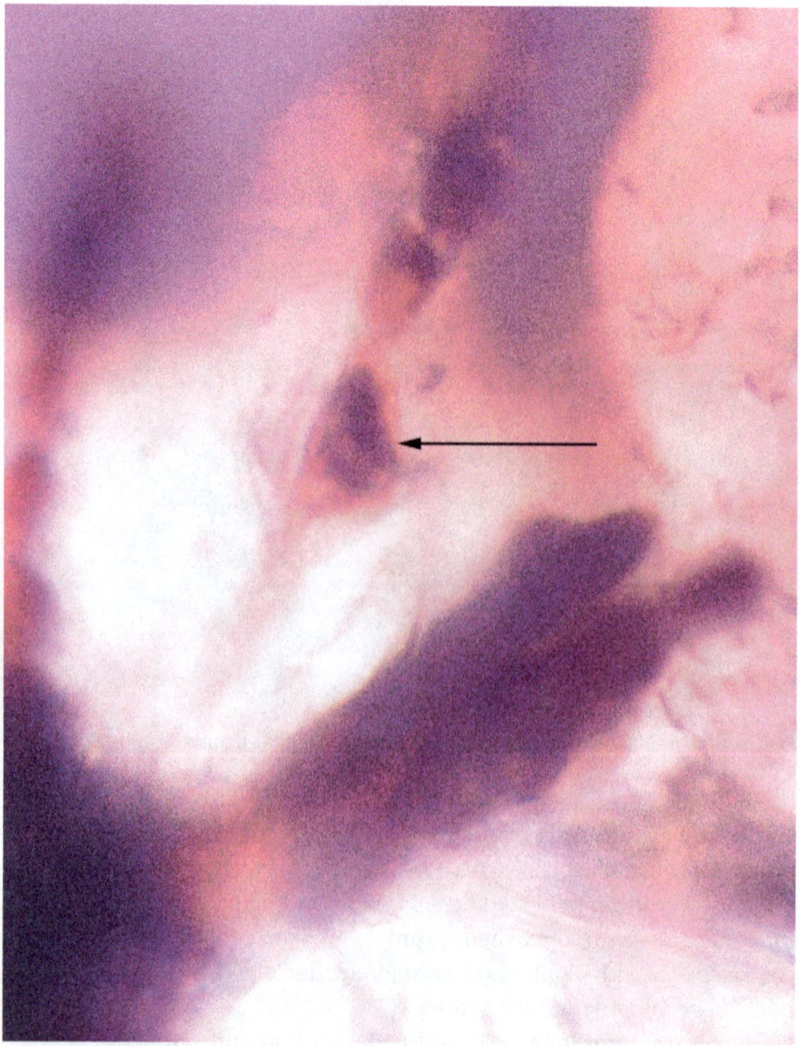

Figure 7. *Devalquia brisaci*, holotype specimen IGR.PLC-1, photograph of maxillary palps (arrow indicates broadened third palpomere).

Abdomen 0.46 mm long (without genitalia), 0.17 mm wide; genitalia not visible.

Discussion

Devalquia brisaci gen.n., sp.n. has two synapomorphies of the (Forcipomyiinae + Dasyheleinae + Ceratopogoninae) (proposed by Borkent et al.

1987: 595, 599), viz., trochanter of fore and mid legs with a pair of thick, contiguous setae; anapleural suture short, extending to posterior margin of anepisternal cleft. A phylogenetic analysis of the relationships between these subfamilies is long overdue. *Devalquia* shares with the subfamily Forcipomyiinae the following characters (after Downes & Wirth 1981): wing with crossvein r-m present; female with 13 flagellomeres; wing with numerous macrotrichia. Nevertheless, *Devalquia* has no empodia and its claws are gently curved, which are not characters of this subfamily, but better correspond to *Dasyhelea*, Kieffer, 1911 (Dasyheleinae), together with the presence of numerous macrotrichia on wings, as in *Forcipomyia*. Chan & Leroux (1971: 1324) noted that "Difficulty in distinguishing adult females of *Dasyhelea* and *Lasiohelea*, Kieffer 1921 (a *Forcipomyia* subgenus) species has often arisen from their close morphological similarity, especially in wing and antennal characters". Lee & Reye (1954: 244) have the same opinion, adding that *Dasyhelea* can be separated from *Lasiohelea* on the basis of the absence of an empodium.

Therefore, we have to discuss the affinities of *Devalquia* with both Forcipomyiinae and Dasyheleinae. The Forcipomyiinae comprise the three genera *Atrichopogon* Kieffer, 1906, *Atriculicoides* Remm, 1976 and *Forcipomyia* Meigen, 1818. All these taxa are clearly difficult to separate. Szadziewski (1996: 51) treated the Dasyheleini and Forcipomyiini as tribes within the Forcipomyiinae, with a third tribe Atriculicoidini Szadziewski 1996 for *Atriculicoides*.

On one side, the Dasyheleinae only comprise *Dasyhelea*, the unique difference between *Devalquia* and the latter being the absence of sculptured flagellomeres in the fossil. On the other side, following the keys of Downes & Wirth (1981), *Devalquia* would fall near *Forcipomyia* rather than near an *Atrichopogon* because it has minute microtrichia and macrotrichia covering most of the wing, fringe complex (not a single row of hairs), cell R2+3 rather short. Nevertheless Borkent & Craig (2004: 59) indicated that they have been unable to identify a synapomorphy for *Forcipomyia*, which could be paraphyletic in relation to *Atrichopogon*.

When compared with the *Forcipomyia* subgenera, *Devalquia* is more likely to correspond to the subgenus *Lasiohelea* for the presence of a sensory pit on the third palpal segment; proximal flagellomeres transverse to spherical, much shorter than distal five; second radial cell long and narrow; CR 0.65 (between 0.54 and 0.70) and TR(III) 2.5 (between 1.0 and 3.0). Nevertheless *Devalquia* has no empodia unlike all the Recent *Lasiohelea* species. Differences with the Cenozoic species *Lasiohelea woodruffi* Szadzewski & Grogan, 1998, *Lasiohelea americana* Szadzewski

& Grogan, 1998, and *Lasiohelea succinea* Szadzewski, 1988, are as follows: the presence of a sensory pit on third palpal segment excludes affinities with *Lasiohelea woodruffi*. *Lasiohelea americana* has broad small claws which excludes affinities with *Devalquia* that has slender claws. *Lasiohelea succinea* has long claws, total body length 1.5 mm and CR 0.7 which excludes affinities with *Devalquia* (Szadzewski 1988; Szadzewski & Grogan 1998).

Devalquia shares with the Cretaceous genus *Atriculicoides* the presence of eyes broadly contiguous, a terminal flagellomere rounded at apex and a hind tibial spur (Borkent 1995), identical wing venation and ornamentation. The female claws with distinctly bifid apices could be a difference with some species of *Atriculicoides* but this character is not well known in this genus. *Devalquia* differs from *Atriculicoides swinhoei* (Cockerell, 1919) (Burmese amber) in the wing membrane with numerous macrotrichia (Szadziewski 2004; Szadziewski & Poinar 2005); it differs from *Atriculicoides szadziewskii* Perez-de La Fuente et al., 2011, based on a female from the Albian of Spain, in the macrotrichia covering on the basal radial cell plus the distal half of the wing membrane (also covering median and cubital areas in *Devalquia*) (Perez-de La Fuente et al. 2011). *Atriculicoides incompletus* Szadziewski & Schlüter, 1992 has four-segmented palps, unlike *Devalquia* (Szadziewski & Schlüter, 1992). The first tarsomere of foreleg without row of stout bristles exclude affinities with *Atriculicoides globosus* (Boesel, 1937) (Borkent 1995). The posterior wing fringe with slender setae would suggest affinities with *Atriculicoides macrophthalmus* Remm, 1976. But the species based on male only (viz., *Atriculicoides cenomanensis* Szadziewski & Schlüter, 1992) are hardly comparable with *Devalquia*. It remains that Szadziewski (1996) erected a tribe Atriculicoidini for the genus *Atriculicoides*, with the female claws with distinctly bifid apices as diagnostic character. *Devalquia* has clearly simple apices of claws, excluding it from this tribe and from *Atriculicoides*. Nevertheless the present taxonomic situation is confuse because, for instance *Atriculicoides szadziewskii* is supposed to have simple claws (Perez-de La Fuente et al. 2011).

In conclusion, if *Devalquia* is clearly related to the Dasyheleinae and Forcipomyiinae, its exact affinities within these groups remain uncertain. It does not correspond to any of the Recent or fossil genera. Thus we choose to consider that it represents a new genus of uncertain affinities in the (Dasyheleinae & Forcipomyiinae). A phylogenetic analysis of these midges is necessary to precise its affinities.

Subfamily Ceratopogoninae Newmann, 1834
Tribe Heteromyiini Wirth, 1962
Genus *Metahelea* Edwards. 1929

***Metahelea roggeroi* Choufani & Nel sp. n.** (Figs 8–16)

Type specimen

Holotype specimen IGR.PLC-2 (female No. AO 7 in Daniel Roggero coll.), deposited in the amber collection of the Geology Department and Museum of the University Rennes 1, France.

Figure 8. *Metahelea roggeroi* Choufani & Nel sp. n., holotype specimen AO 7 IGR.PLC-2, photograph of general habitus.

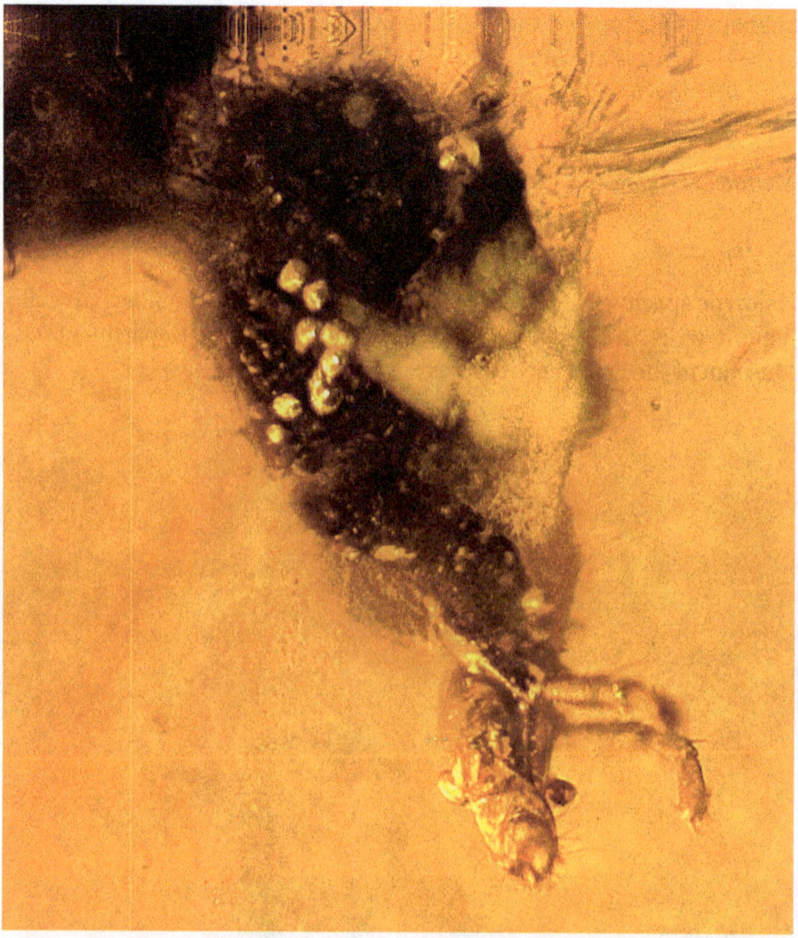

Figure 9. *Metahelea roggeroi*, holotype specimen IGR.PLC-2, photograph of lateral view of head showing palps and proboscis.

Diagnosis

Fifth tarsomere of fore leg swollen; presence of an acute and very long spine on dorsal arm of each lobe of fourth tarsomere; wings hyaline.

Age and locality

Early Santonian, sandy-to-coaly clayey beds (geological stratum 'a' in Gomez et al. 2003: fig. 1) in ancient lignite mine of Massif d'Uchaux, Piolenc, Vaucluse, France.

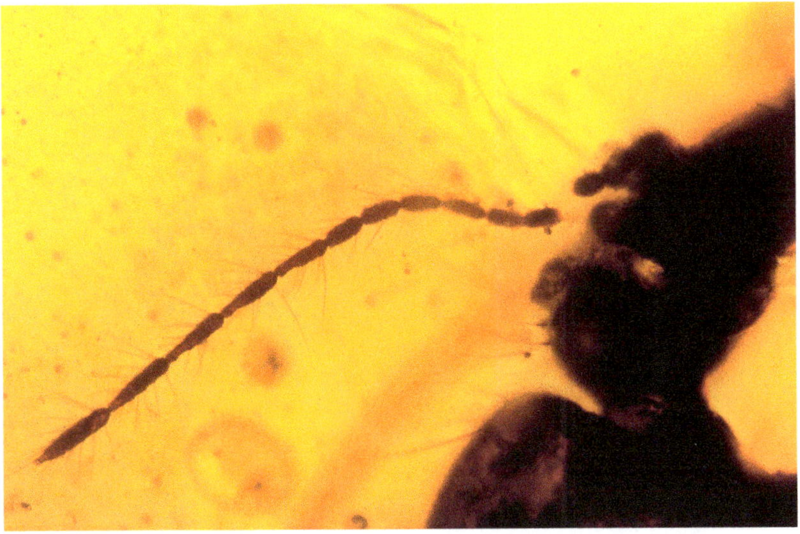

Figure 10. *Metahelea roggeroi*, holotype specimen IGR.PLC-2, photograph of antennae.

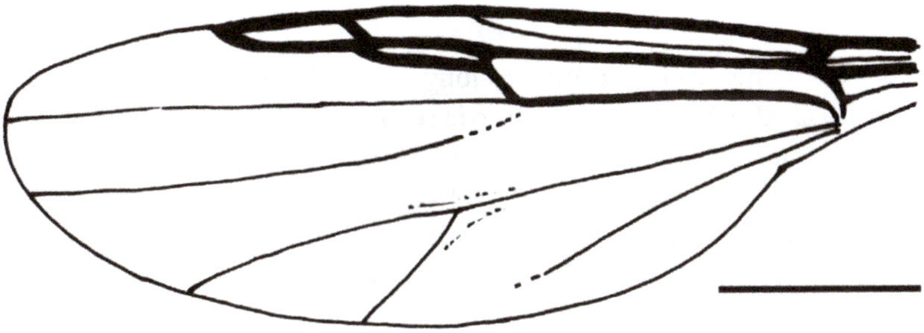

Figure 11. *Metahelea roggeroi*, holotype specimen AO 7, drawing of wing. Scale bar=0.2 mm.

Etymology

Named after our friend and colleague Daniel Roggero who discovered the type specimen.

Description

Female body circa 1.0 mm long; head black; proboscis length 1.2 times as height of head; maxillary palpi dark brown, RL-P 0.16 : 0.17 : 0.4 : 0.07 : 0.2, PR 2.67; size of antennal scape not determinable, antennal pedicel 0.045 mm long and 0.052 mm wide, flagellum distinctly annulate, 13 flagellomeres,

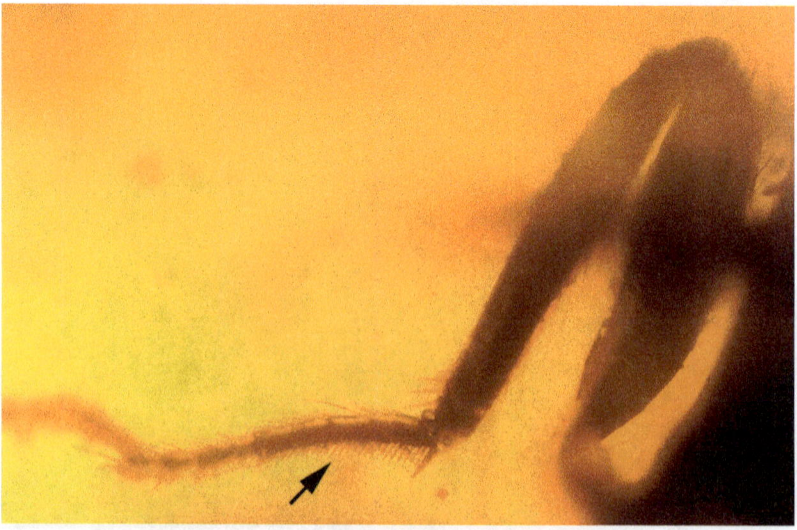

Figure 12. *Metahelea roggeroi*, holotype specimen IGR.PLC-2, photograph of hind leg showing tarsal ctenidia (arrow).

short basal ones each 2.14 times as long as broad, distal five flagellomeres cylindrical, RL-8A 0.08 : 0.08 : 0.09 : 0.12 : 0.14 : 0.14 : 0.165 : 0.185, AR 1.17; eyes touching medially; proboscis moderately long, palpus five-segmented, distal segments cylindrical, third palpal segment bearing 3–4 long sensilla capitata; fourth segment armed with two setae; fifth segment elongate.

Thorax: scutum with three pairs of long setae in anterior part and one pair of prescutellar setae, scutellum with two pairs of setae, D-shaped anteroanepisternum and a narrow anapleural cleft; a broad triangular paratergite; fore femur 0.32 mm long, 0.076 mm wide, not swollen, with four lateral long setae, fore tibia not swollen, 0.34 mm long, 0.064 mm wide, with two series of short, straight and strong setae, ctenidia-like; mid femur 0.3 mm long, 0.056 mm wide, mid tibia 0.55 mm long, 0.048 mm wide, both with two series of sparse but rather long setae; hind femur 0.43 mm long, 0.09 mm wide, hind tibia 0.45 mm long, 0.067 mm wide, both with series of long but weak setae, a strong apical tibial spur and an apical ctenidia formed by four short, straight and strong setae; hind leg with 1st tarsomere bearing distinct ctenidia, 2nd tarsomere 3/4 the length of 1st tarsomere; 4th tarsomere of all legs deeply bilobed, each lobe further subdivided into two subsidiary lobes, the lower of which bears a strong, blunt apical spine, and the upper bears a strong straight and longer spine;

Figure 13. *Metahelea roggeroi*, holotype specimen IGR.PLC-2, photograph of hind leg showing apical claw and deeply bilobed fourth tarsomere.

5th tarsomeres of all legs with distinct ventral swelling covered with fine setae at midlength; 5th fore tarsomere very similar to that of *Tetrabezzia* Kieffer 1917 (Clastrier 1982: fig. 1K), ratio of width at midlength of 5th tarsomere of foreleg to width at midlength of 5th tarsomere of midleg = 2.0; all claws single, slightly longer than 5th tarsomere, each with distinct short basal tooth; claw of all legs 0.94 the length of 5th tarsomere, basal tooth 0.37 times length of claw.

Wings hyaline, 0.475 mm long, 0.26 mm wide; costa not extended beyond apex of R3; cells R1 short quadrangular, 0.87 times length of R2+3; R2+3 not squared-ended, acute at apex, ending well beyond middle of wing; interruption of vein r-m (most likely a plesiomorphic feature, present in *Neurohelea*, after Tóthová et al. 2009); M1 and M2 apparently forking just basal to r-m but not well preserved; wing membrane without macrotrichia.

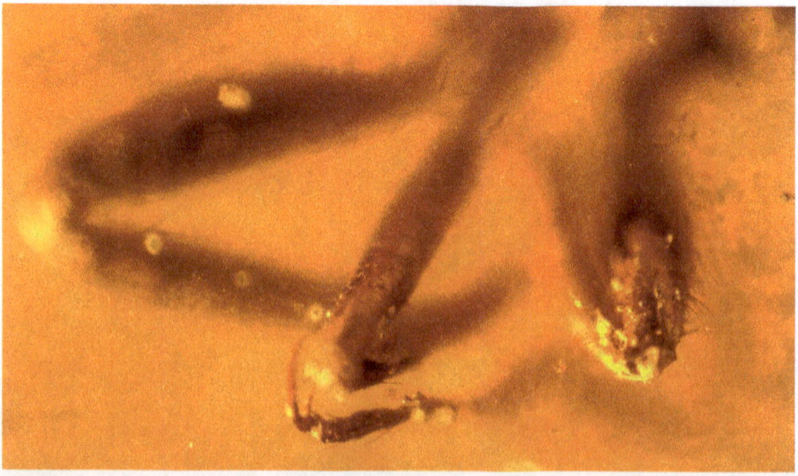

Figure 14. *Metahelea roggeroi*, holotype specimen IGR.PLC-2, photograph of fore leg showing apical claw.

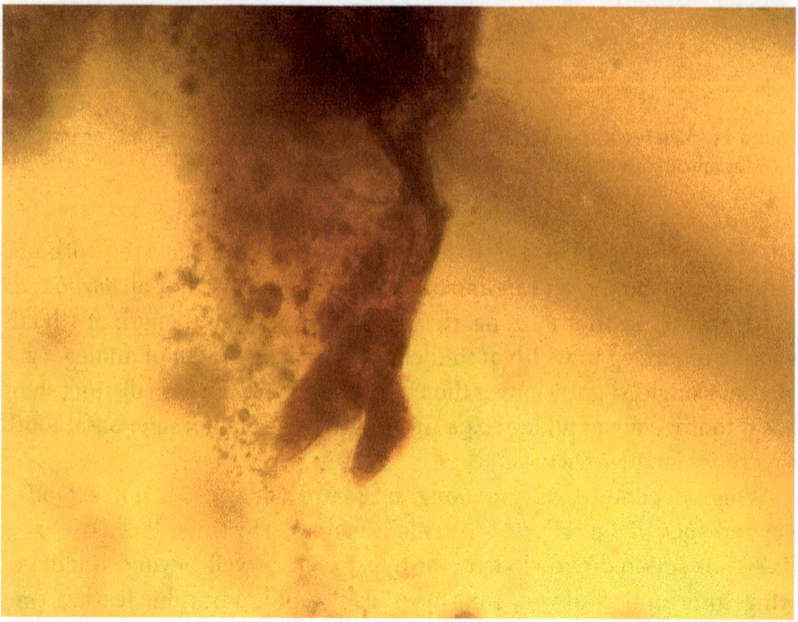

Figure 15. *Metahelea roggeroi*, holotype specimen IGR.PLC-2, photograph of female apex of abdomen.

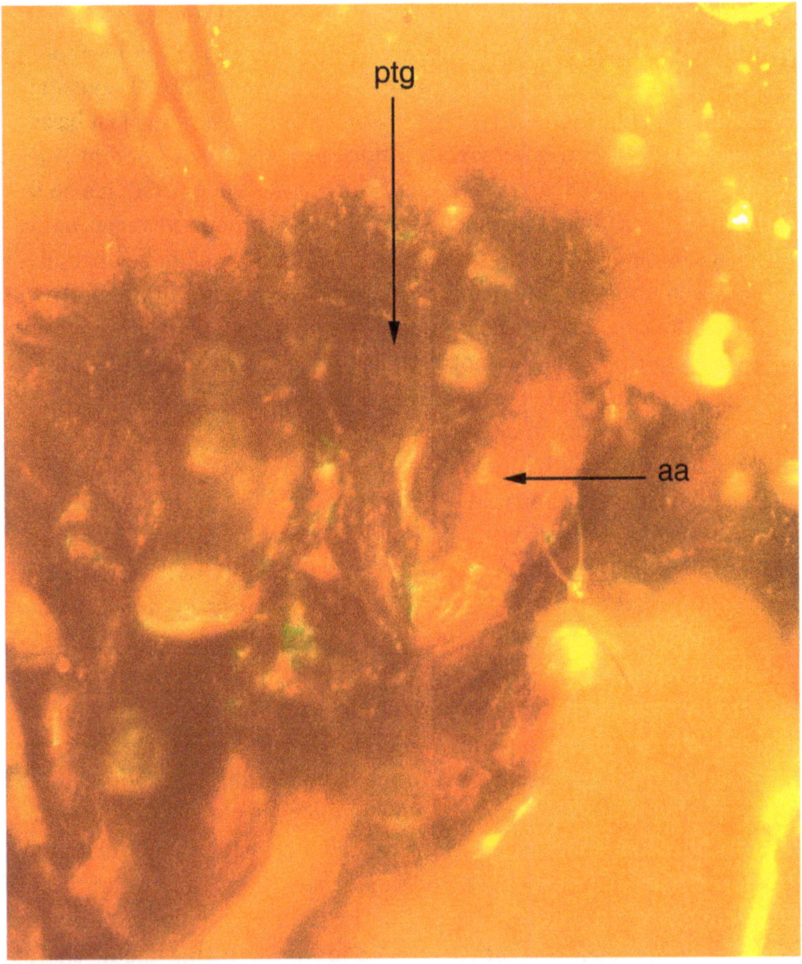

Figure 16. *Metahelea roggeroi*, holotype specimen IGR.PLC-2, photograph of lateral side of thorax. Abbreviations: aa, anteroanepisternum; ptg, paratergite.

Abdomen in poor condition, deformed, 0.38 mm long, with two visible cerci, 0.05 mm long; genitalia not visible.

Discussion

Metahelea roggeroi sp. n. has a broad triangular paratergite, D-shaped anteroanepisternum and a narrow anapleural cleft, characteristic of the Heteromyiini, Sphaeromiini, Palpomyiini, and Stenoxenini (after

Kaczorowska 2000). *Metahelea roggeroi* can be attributed to the Heteromyiini on the basis of the following characters (after Borkent 1998): swollen 5th tarsomere on foreleg (synapomorphy), midleg 4th tarsomere markedly bifid and spinose, plus probably M1 and M2 forking just basal to r-m, lacking petiolate base, viz., apomorphic state 53′ of Borkent (1995). After the key to the Nearctic genera of Downes & Wirth (1981), it also falls in the Heteromyiini for the following characters: empodia vestigial, not visible; flagellomeres not sculptured; cells R1 and R2+3 well-developed; R2+3 not squared-ended, acute at apex, ending well beyond middle of wing; maybe M forking just before r-m; 5th tarsomeres of female armed with slender sharp-tipped spines, not blunt spines; claws of female unequal.

After the key to the heteromyiine genera of Grogan & Wirth (1979), the new taxon falls within *Metahelea* for the following characters: M forking just proximal to r-m; 4th hind tarsomere bifid, spinose; 5th hind tarsomere as long as those of fore and mid legs; hind claw not greatly elongate; "claws" of fore leg unequal (in fact it is a long claw with a shorter basal tooth). Borkent (1998) has modified and proposed a new key, but it presents a difficulty at the node '7' (concerning the females) because this author characterizes the first part of the dichotomy (leading to node '8') by 'hind 5th tarsomere with single, extremely elongate claw', while the second part of dichotomy (leading to node '10') would be characterised by 'hind 5th tarsomere with two claws'; but first part of the dichotomy in node '10' is characterized by 'mid and hindleg with one claw bearing one shorter tooth', which is obviously contradictory with node '7', furthermore subsequent node '11' goes to *Metahelea*, which has also foreleg with a single claw.

Therefore we have to explore for *Metahelea roggeroi* the two options of the dichotomy of node '7'. First option is leading to node '8' and would send *Metahelea roggeroi* near *Tetrabezzia* for the following female characters: hind 5th tarsomere with single claw, with much shorter basal tooth; hind leg with 2nd tarsomere 3/4 the length of 1st tarsomere; foreleg not swollen, with at most a few thick spines on tibia; 4th tarsomeres of fore and midleg bifid and spinose. However, the group (*Heteromyia* Say, 1825, *Pellucidomyia* Macfie, 1939, *Tetrabezzia*) is characterised by a set of characters absent in *Metahelea roggeroi*, viz., hind claw extremely elongate, hind leg strikingly elongate, and greatly elongate hind 4th and 5th tarsomeres and claws (a synapomorphy of this group after Grogan & Wirth 1979) (direct observation and see Wirth & Ratanaworabhan 1971). Lastly the cell R1 is absent in this group while it is present in *Metahelea roggeroi*.

The second option is leading to node '10' and subsequently to *Metahelea* for the following characters: mid and hindleg with one claw bearing one shorter tooth; 4th tarsomeres of midleg bifid and spinose; 4th tarsomeres of hindleg bifid and spinose, synapomorphy of the group of genera (*Ceratobezzia* Kieffer, 1917, treated by Borkent (2006) as a subgenus of *Clinohelea* Kieffer, 1917, *Metahelea* and *Clinohelea*), after Grogan & Wirth (1979); fore leg with a single claw with a basal tooth; 4th fore tarsomere not cordiform and with spines; 4th tarsomere of fore leg deeply bilobed, each lobe further subdivided into two subsidiary lobes. *Metahelea roggeroi* would differ from the three other species of *Metahelea* in the 5th tarsomere of fore leg more swollen. Szadziewski (1998) indicated that the 5th tarsomeres are similar on all legs in the Baltic amber *Metahelea serafini* Szadziewski 1998. A further difference with *M. serafini* and the two recent species *M. bifasciata* Kieffer 1917 and *M. metallescens* Edwards 1929 is the presence of an acute and very long spine on dorsal arm of each lobe of 4th tarsomere (Edwards 1929; Debenham 1974; Szadziewski 1998). *M. bifasciata* and *M. metallescens* have colored wings; it is also probably the case for *M. serafini*, but not for *Metahelea roggeroi*.

There is a further small difficulty in the key of Borkent (1998), i.e. it is indicated that the eyes are contiguous in *Clinohelea* while they should be broadly separated medially in *Metahelea* and *Neurohelea* Kieffer, 1925; but in the key of Grogan and Wirth (1979), it is exactly the contrary (eyes contiguous in *Metahelea* while separated in *Clinohelea*). Edwards (1929) wrote that the eyes are touching in *Metahelea*, Debenham (1974) also indicated that the eyes are very closely contiguous in *Metahelea*, while Clastrier (1983) added that they are very widely separated in *Clinohelea*, and Tóthová et al. (2009) figured a *Neurohelea* with eyes well separated. In *Metahelea roggeroi*, the eyes are touching like in the other *Metahelea* species but it is difficult to determine if it is for a long distance of not.

Conclusion

Devalquia gen. n. is the second Cretaceous Forcipomyiinae & Dasyheleinae, after *Atriculicoides* (Szadziewski 1996). Its strong similarities with *Dasyhelea* and *Lasiohelea* suggest that it could represent a stem group of these taxa.

Beckenbach & Borkent (2003: 31, fig. 4) indicated that the Heteromyiini belong to a ceratopogonine clade that also comprises Sphaeromiini, Palpomyiini, and Stenoxenini. The sphaeromiine genus *Probezzia* Kieffer, 1906 is the unique representative of this clade that they included in their molecular phylogenetic analysis of the Ceratopogonidae. Despite the incomplete information on the phylogenetic relationships within the Ceratopogoninae, it is possible to infer from the terminal position of *Probezzia* in this phylogeny that the clade (Heteromyiini, Sphaeromiini, Palpomyiini, Stenoxenini) is among the most derived ones in the Ceratopogoninae. Up to now only three fossil Heteromyiini have been described, all from the Cenozoic, viz., *Heteromyia dominicana* Szadziewski & Grogan, 1997 (Miocene, Dominican amber), *Metahelea serafini* Szadziewski, 1998 (Eocene, Baltic amber) and *Physohelea obtuse* Meunier, 1904 (Eocene, Baltic amber). The Baltic amber *Mallochohelea martae* Szadziewski, 2005 is the unique fossil Sphaeromiini. The fossil Palpomyiini are *Palpomyia erikae* Szadziewski, 1993 (Miocene, Germany), *P. edwardsi* Cockerell, 1921 (Late Eocene, Great Britain), *P. freyi* Pierce, 1966, *P. multispinosa* Pierce, 1966, *P. shilo* Pierce 1966, *P. ryshkoffi* Pierce, 1966 (all Miocene, California), *P. jantari* Szadziewski, 1988, *P. riedeli* Szadziewski, 1988, *P. succinea* Szadziewski, 1988 (all Eocene, Baltic amber), *P. wirthorum* Szadziewski & Grogan, 1997 and *Phaenobezzia wirthi* Szadziewski & Grogan, 1997 (all Miocene, Dominican amber). No fossil Stenoxenini is known. Lastly the Eocene Fushun amber taxon *Eopalpomyitis unca* Hong, 2002 is currently considered as an unplaced (Heteromyiini + Sphaeromiini + Palpomyiini + Stenoxenini). Therefore, *Metahelea roggeroi* is the first Mesozoic and earliest record of this clade. It suggests that the other ceratopogonine clades were also present as early as in the Upper Cretaceous, contra the opinion of Szadziewski (1996: 77, fig. 46) based on the available data by that time. Its attribution to a Recent genus confirms the great morphological stability through time within this family (Borkent 2001; Borkent & Craig 2004), clearly more important than in its sister group Chironomidae. In this last family, if nearly all the Recent subfamilies are also known in the Cretaceous, it is not the case for the modern genera (Azar & Nel 2012). A possible reason could be the great changes in the diet that took place in chironomids from hematophagy to nectarivory during the Cretaceous, likely linked to the diversification of the angiosperms. Meantime, the Ceratopogonidae have not drastically changed their diet of hematophagy and/or insectivory.

Another question is the highly specialized leg structures of the Heteromyiini, clearly developed in *Metahelea roggeroi*, viz., presence of a unique very strong claw with a strong and shorter basal tooth, plus that of series of ctenidia on tibiae and tarsi. Such structures are strikingly similar to those developed in ectoparasite insects (Siphonaptera, "Phthiraptera", Diptera of Hippoboscidae or Streblidae, etc.) for adaptation to stay on their hosts. Such structures are evenly present in giant

Mesozoic fleas recently described (Huang et al. 2012). The Recent Heteromyiini are supposed to be insectivorous (at least some of them), with a very particular mating behavior where the female is grasping the male from above and eats it during copulation (Debenham 1974). Possibly the specialized leg structures of the Heteromyiini were developed to favour a better grasping of the female on the male or on other insects, but maybe were originally an adaptation to ectoparasitism on vertebrates or arthropods, as already indicated by Lane (1977) for the claws and ctenidia of females of species of *Forcipomyia* (*Trichohelea* Goetghebuer, 1920) as adaptations to cling to the scales of butterfly wings as females feed on butterfly blood.

Acknowledgements

We sincerely thank Christian Devalque, Daniel Roggero and Patrick Brisac for their great help in the re-discovery of Piolenc outcrop and the field studies. This is a contribution to the team project "Biodiversity: Origin, Structure, Evolution and Geology" awarded to D.A. by the Lebanese University. Field work in July 2012 was supported by the program Interrvie (Research Project NOVAMBRE 2 to D. Néraudeau, Univ. Rennes 1) from the French National Institute for Universe Sciences (CNRS-INSU).

References

Azar, D. & Nel, A. (2012) Evolution of hematophagy in "non-biting midges" (Diptera: Chironomidae). *Terrestrial Arthropod Reviews* 5: 15–34.

Azar, D., Gèze, R., El-Samrani, A., Maalouly, J. & Nel, A. (2010) Jurassic Amber in Lebanon. *Acta Geologica Sinica* (English Edition) **84**: 977–983.

Azar, D., Perrichot, V., Néraudeau, D. & Nel, A. (2003) New psychodid flies from the Cretaceous ambers of Lebanon and France, with a discussion about *Eophlebotomus connectens* Cockerell, 1920 (Diptera, Psychodidae). *Annals of the Entomological Society of America* **96**: 117–127.

Beckenbach, A.T. & Borkent, A. (2003) Molecular analysis of the biting midges (Diptera: Ceratopogonidae), based on mitochondrial cytochrome oxidase subunit 2. *Molecular Phylogenetics and Evolution* **27**: 21–35.

Borkent, A. (1998) A revision of *Neurobezzia* Wirth and Ratanaworabhan and *Neurohelea* Kieffer, with a description of a new genus and discussion of their phylogenetic relationships (Diptera: Ceratopogonidae). *Entomologica Scandinavica* **29**: 137–160.

Borkent, A. (2000) Biting midges (Ceratopogonidae: Diptera) from Lower Cretaceous Lebanese amber with a discussion of the diversity and patterns found in other ambers. In: Grimaldi, D. (Ed.) *Studies on fossils in amber, with particular reference to the Cretaceous of New Jersey*. Backhuys, Leiden, pp. 355–451.

Borkent, A. (2001) *Leptoconops* (Diptera: Ceratopogonidae), the earliest extant lineage of biting midge, discovered in 120–122 million-year-old Lebanese amber. *American Museum Novitates* **3328**: 1–11.

Borkent, A. (2006) *World species of biting midges (Diptera: Ceratopogonidae)*. Available online at http://www.inhs.uiuc.edu/cee/FLYTREE/CeratopogonidaeCatalog.pdf.

Borkent, A. & Craig, D.A. (2004) *Austroconops* Wirth and Lee, a Lower Cretaceous genus of biting midges yet living in Western Australia: a new species, first description of the immatures and discussion of their biology and phylogeny (Diptera: Ceratopogonidae). *American Museum Novitates* **3449**: 1–67.

Borkent, A., Wirth, W.W. & Dyce, A.L. (1987) The newly discovered male of *Austroconops* (Ceratopogonidae: Diptera) with a discussion of the phylogeny of the basal lineages of the Ceratopogonidae. *Proceedings of the Entomological Society of Washington* **89**: 587–606.

Chan, K.L. and Le Roux, E.J. (1971) A new subgenus and species of midge (Diptera: Ceratopogonidae) from Singapore. *The Canadian Entomologist* **103**: 271–276.

Clastrier, J. (1983) Description de deux nouvelles especes afrotropicales concernant les genres *Ceratobezzia* et *Clinohelea* (Diptera, Ceratopogonidae). *Revue Francaise d'Entomologie* (N.S.) **5**: 19–26.

Debenham, M.L. (1974) A revision of the Australian and New Guinea predatory Ceratopogonidae (Diptera: Nematocera) of the tribes Heteromyiini and Sphaeromiini. *Australian Journal of Zoology* (Supplementary Series) **28**: 1–92.

Downes, J.A. & Wirth, W.W. (1981) Ceratopogonidae, p. 393–421. In: McAlpine, J.F., Peterson, B.V., Shewell, G.E., Teskey, H.J., Vockeroth, J.R. and Wood, D.M. (Coordinators), *Monograph 27: Manual of Nearctic Diptera. Volume 1*. Research Branch Agriculture Canada, Ottawa, ON, vi + 674 pp.

Dumas, E. (1876) *Statistique géologique, minéralogique, métallurgique et paléontologique du département du Gard*. Paris, Nimes, Alais, 3 volumes: 274 pp. + 723 pp. + 511 pp.

Edwards, F.W. (1929) Philippine nematocerous Diptera II. *Notulae Entomologicae* **9**: 1–14.

Gomez, B., Barale, G., Saad, D. & Perrichot, V. (2003) Santonian angiosperm-dominated leaf-assemblage from Piolenc (Vaucluse, Sud-Est de la France). *Comptes Rendus Palevol* **2**: 197–204.

Grogan, W.L. & Wirth, W.W. (1979) A new Neotropical genus of predaceous midges, with a key to the genera of Heteromyini (Diptera: Ceratopogonidae). *Proceedings of the Entomological Society of Washington* **81**: 51–59.

Kaczorowska, E. (2000) The thoracic morphology of biting midges (Diptera: Ceratopogonidae). *Polish Journal of Entomology* **69**: 87–131.

Lacroix, A. (1910) Résines fossiles. pp. In: Lacroix, A. (Ed.) *Minéralogie de la France, Volume 4*. Paris, pp. 637–645.

Lane, R.P. (1977) Ectoparasitic adaptations in *Forcipomyia* from butterflies with two new African species (Ceratopogonidae). *Systematic Entomology* **2**: 305–312.

Lee, D.J. & Reye, E.J. (1954) Australasian Ceratopogonidae (Diptera, Nematocera). Part VII. Notes on the Genera *Alluaudomyia*, *Ceratopogon*, *Culicoides* and *Lasiohelea*. *Proceedings of the Linnean Society of New South Wales* **79**: 233–246.

Onoratini, G., Guiliano, M., Mille, G. & Simon, P. (2009) L'ambre albo-cénomanien de la montagne de Lure (Alpes-de-Haute-Provence), outil stratigraphique et paléogéographique. *Geobios* **42**: 89–99.

Remm, C.J. (1976) Late Cretaceous biting midges (Diptera: Ceratopogonidae) from fossil resins of the Khatanga depression. *Paleontological Journal* **10**: 344–351.

Schmidt, A.R., Jancke, S., Lindquist, E.E., Ragazzi, E., Roghi, G., Nascimbene, P.C., Schmidt, K., Wappler, T. & Grimaldi, D.A. (2012) Arthropods in amber from the Triassic period. *Proceedings of the National Academy of Sciences of the United States of America* **109**: 14796–14801.

Szadziewski, R. (1988) Biting midges (Diptera: Ceratopogonidae) from Baltic amber. *Polskie Pismo Entomologiczne* **58**: 3–283.

Szadziewski, R. (1996) Biting midges from Lower Cretaceous amber of Lebanon and Upper Cretaceous Siberian amber of Taimyr (Diptera: Ceratopogonidae). *Studia Dipterologica* **3**: 23–86.

Szadziewski, R. (1998) A new species of predaceous midge genus *Metahelea* from Baltic amber (Diptera: Ceratopogonidae). *Polskie Pismo Entomologiczne* **67**: 245–254.

Tóthová, A., Knoz, J. & Gelbic, I. (2009) On morphology and distribution of *Neurohelea luteitarsis* (Wald) (Diptera: Ceratopogonidae) in Europe. *Annales de la Société Entomologique de France* (N.S.) **45**: 297–301.

PART THREE

INSECTS FROM LOWER CRETACEOUS LEBANESE AMBER

Gapenus rhinariatus gen. sp. n., a new whitefly from Lebanese amber (Hemiptera: Sternorrhyncha: Aleyrodidae)

Jowita Drohojowska[a] and Jacek Szwedo[b,*]

[a]Department of Zoology, University of Silesia,
9 Bankowa Street, PL40-007 Katowice, Poland
[b]Department of Palaeozoology, Museum and Institute of Zoology,
Polish Academy of Sciences, 64 Wilcza Street, PL00-679 Warszawa, Poland
*Corresponding author, e-mail: szwedo@miiz.waw.pl

Abstract

Gapenus rhinariatus gen. et sp. n. from the Lower Cretaceous Lebanese amber is described, based on an adult male specimen. It is the second representative of subfamily Aleurodicinae (Hemiptera: Sternorrhyncha: Aleyrodidae) in the fossil record and the oldest representative of this subfamily known so far. Morphological features of this fossil are discussed.

Keywords

Gapenus gen. n.; *G. rhinariatus* sp. n.; new genus; new species; fossil; morphology; taxonomy; evolution

Introduction

The suborder Sternorrhyncha (Hemiptera) comprises a number of evolutionary lineages: Psylliformes (*sensu* Schlee 1969a), i.e., Psyllomorpha + Aleyrodomorpha, and Aphidiformes (*sensu* Schlee 1969b), i.e., Pincombeomorpha and their descendants Coccidomorpha and Aphidomorpha (Drohojowska & Szwedo 2011a). The monophyly of the whole Sternorrhyncha, as well as status of its particular units is still under discussion (Schlee 1969c; Hennig 1981; Shcherbakov 1990, 2000, 2005, 2007; Shcherbakov & Popov 2002; Gullan & Martin 2003; Grimaldi & Engel 2005; Forero 2008).

The earliest fossil record of Aleyrodidae (the sole family of Aleyrodomorpha) comes from the Upper Jurassic (Shcherbakov 2000). A few other fossils are reported from the Lower Cretaceous, Upper

Cretaceous, Palaeogene and Neogene (Schlee 1970; Rietschel 1983; Poinar 1992; Shcherbakov 2000; Antoine et al. 2006; Schmidt et al. 2010; Drohojowska & Szwedo 2011a, b) rock deposits and fossil resins but only a few have been formally described.

A review of fossil records of Aleyrodidae was presented by Drohojowska & Szwedo (2011a), and updated with a new fossil from the Baltic amber (Drohojowska & Szwedo 2011b).

The subfamily Bernaeinae is reported from the Upper Jurassic to the Cretaceous/Palaeogene boundary of Asia (Shcherbakov 2000). The oldest fossil representatives of the Aleyrodinae are recorded from the Lower Cretaceous (Schlee 1970; Shcherbakov 2000; Drohojowska & Szwedo 2011a). The only supposed member of the Aleurodicinae is '*Aleurodicus*' *burmiticus* Cockerell, 1919 from the Lower Cretaceous, Aptian Burmese amber (Cockerell 1919). Placement of '*Aleurodicus*' *burmiticus* in Aleurodicinae was suggested by Shcherbakov (2000), challenging Schlee's (1970) opinion, in which Schlee put in doubts its placement in this subfamily. The genus described below is then the oldest record of the subfamily. Undescribed Udamoselinae (a subfamily of uncertain status, see Martin 2007) are mentioned from the Lebanese amber (Shcherbakov 2000).

Materials and Methods

The studied amber of Lebanon is Lower Cretaceous in age, estimated to be 125–135 million years old and assumed to be produced by gymnosperm trees of the families Araucariaceae or Cheirolepidiaceae (Poinar & Milki 2001; Koteja & Azar 2008). The piece of amber containing the inclusion subject of this study has been prepared, i.e., cut, ground, polished and embedded in Canada balsam between two cover slips as proposed by Azar (2000) and Azar et al. (2003). A Nikon MZ1500 stereoscopic microscope and Nikon Microphot-FX (objectives from 4× to 40×) equipped by a camera lucida and changeable direct and transmitted light were used for microscopic examination. Measurements are approximate because of shriveling and possible optical deformations. Photographs have been taken using a Nikon Microphot-FX with a Nikon Eclipse E 600 digital camera and Lucia® software, and adjusted using Combine ZP and Adobe® Photoshop Elements 6.0.

The general lack of descriptions of extant adult whiteflies makes the description of fossil taxa especially challenging, therefore we are following the model propose in earlier papers (Drohojowska & Szwedo 2011a, b).

Systematic Palaeontology

Order Hemiptera Linnaeus, 1758
Suborder Sternorrhyncha Amyot et Serville, 1843
Infraorder Aleyrodomorpha Chou, 1963
Family Aleyrodidae Westwood, 1840
Subfamily Aleurodicinae Quaintance et Baker, 1913

Genus *Gapenus* gen. n.

Type species

Gapenus rhinariatus sp. n.; by present designation and monotypy.

Diagnosis

The new genus differs from the recent representatives of the subfamily by the presence of 10 antennal segments (up to 8 in modern genera); very distinct and numerous rhinaria on antennal segments III to X (single, at least few rhinaria on some of antennal joints among the modern genera); forewing with stem Sc+R forked basad of half of forewing length (stem Sc+R forked apicad of half of forewing length in *'Aleurodicus' burmiticus*); it differs from the extinct *'Aleurodicus' burmiticus* Cockerell, 1919 by the shape of the hindwing, with subrectangular anteroapical and posteroapical angles and straight apical margin (the angles rounded, apical part rounded in *'Aleurodicus' burmiticus*); hind wing stem Sc+R arcuate (straight in *'Aleurodicus' burmiticus*); branch R1 reaching costal margin distinctly basad of anteroapical angle (branch R1 not reaching margin, directed to anteroapical angle in *'Aleurodicus' burmiticus*); branch Rs slightly curved, reaching margin at anteroapical angle (branch Rs straight directed to apex of hind wing in *'Aleurodicus' burmiticus*).

Remarks

'Aleurodicus' burmiticus Cockerell, 1919 was originally very briefly described, but with a few figures presenting hindwing, antenna, hind leg and male genital appendages. The recent species of the genus *Aleurodicus* Douglas, 1892 were revised by Martin (2008), but the fossil was not included there. In addition, Martin (2008) placed in the genus *Aleurodicus* exclusively the New World species, while for the Old World species he proposed the genus *Palaealeurodicus* Martin, 2008. Mound & Halsey

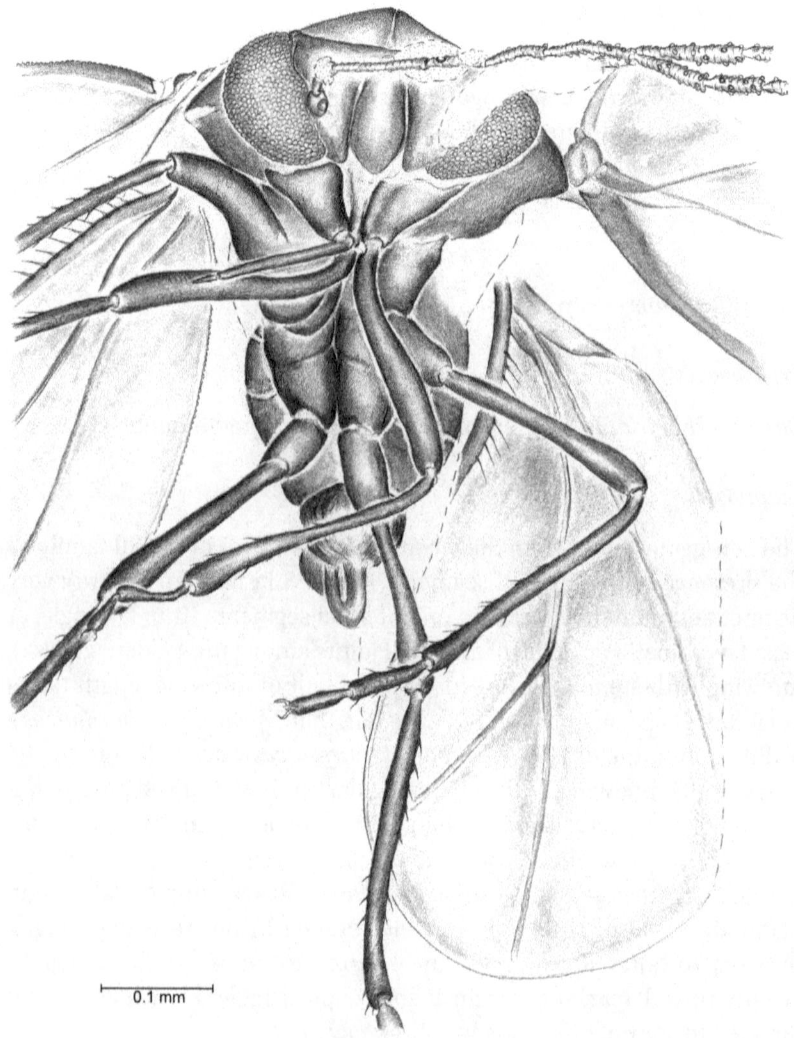

Figure 1. *Gapenus rhinariatus* gen. sp. n. body in anteroventral view.

(1978) pointed out, that fossil Aleyrodidae cannot be placed in the present-day classification. We agree with Shcherbakov (2000) that, '*Aleurodicus*' *burmiticus* Ckll. represent a separate genus, but it is too premature to make any formal decision without detailed re-examination of the specimen.

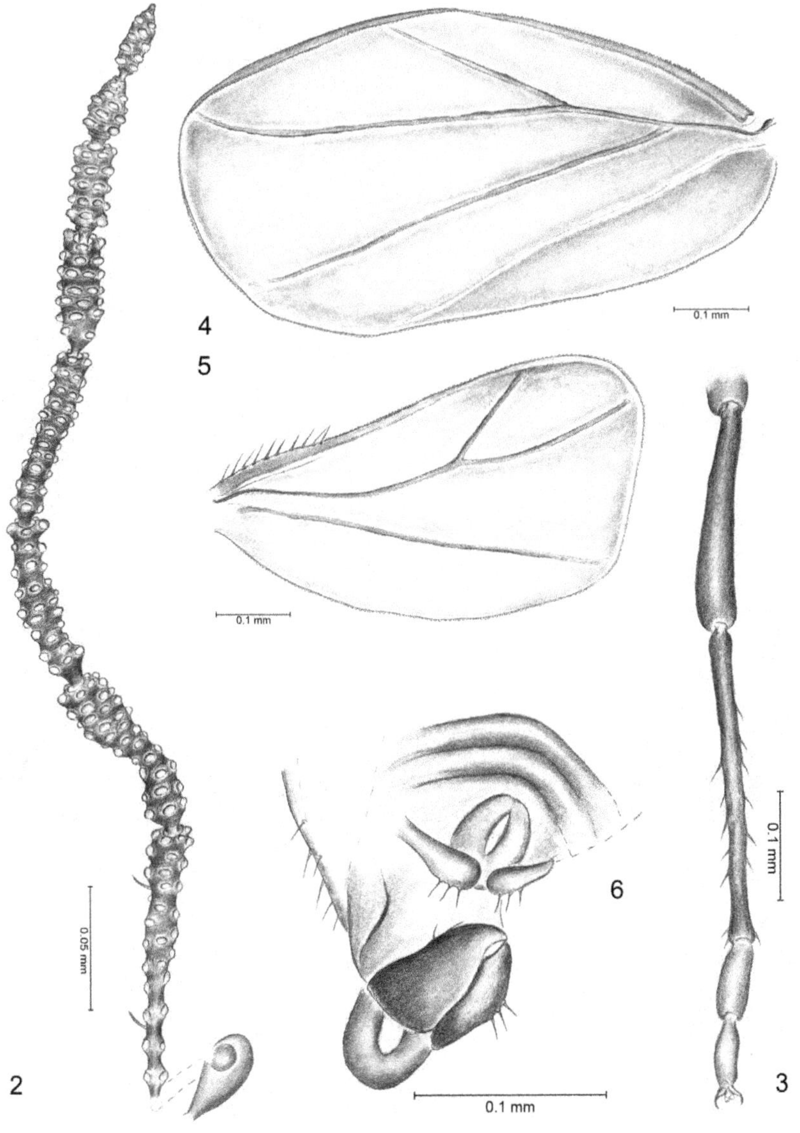

Figures 2–6. *Gapenus rhinariatus* gen. sp. n. (2) Antenna, (3) right hind leg, (4) right forewing, (5) right hind wing, (6) visible portion of male genital block.

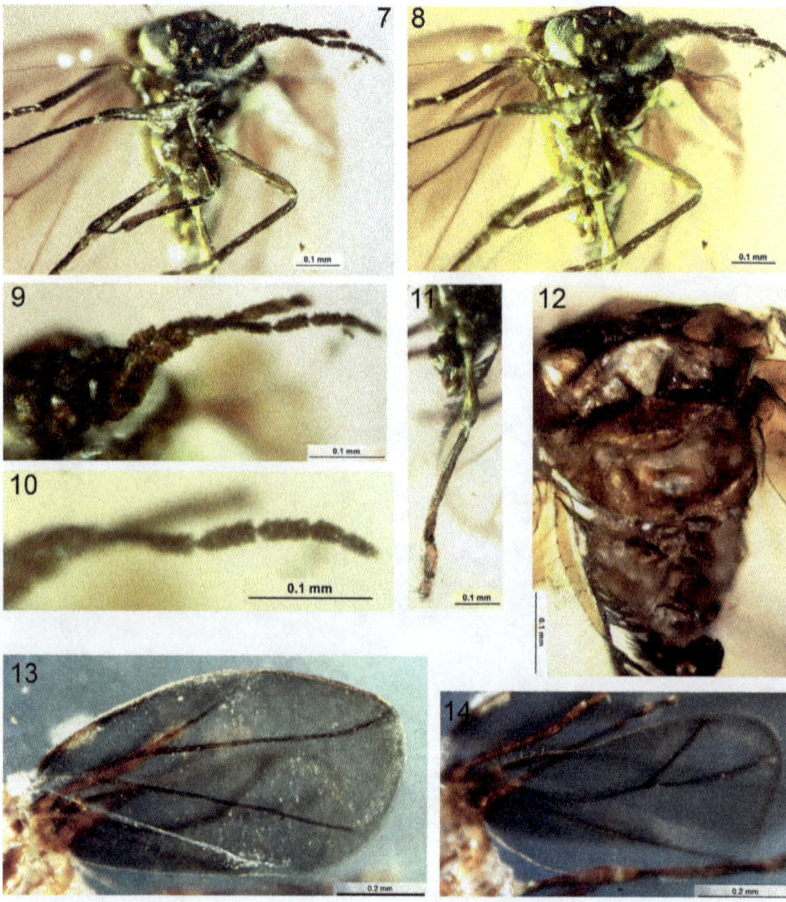

Figures 7–14. *Gapenus rhinariatus* gen. sp. n. (7) Head capsule in anterior view, with basal parts of antennae and fore legs visible, (8) head capsule in anterior view with apical portion of antennae visible, (9) antenna with rhinaria, (10) apical portion of antenna with rhinaria, (11) hind leg, (12) thorax in dorsal view, (13) forewing, (14) hind wing.

Description

Head with compound eyes relatively wide, but narrower than pronotum. Compound eyes not divided. Ocelli placed at anterodorsal angles of compound eyes. Base of antenna in lower half of compound eyes height (in anterior view). Frons about twice as long as wide, widest at basal ⅓; apical portion triangularly converging. Rostrum with apex reaching hind coxae, apical segment short; about 6 times shorter than subapical one. Antennal segment bulbous; antennal segments III the longest; IV shorter, but slightly longer than following antennal segments V and VI, these are equal in length; segments VII to IX shorter each other, segment IX subglobose, tapering apically; segment X slightly longer than preceding one, with acute tip.

Fore legs slightly shorter than mid legs, hind legs the longest; tarsal segments subequal in length. Hind leg with a few sparse setae, not forming distinct combs; tarsal claws not elongated, paronychium lobate.

Forewing about twice as long as wide; with costal margin arcuate, strengthen; anteroapical angle rectangular, apical margin arcuate, posteroapical angle widely arcuate, posterior margin almost straight; apex of forewing at anteroapical angle; margins tuberculate along. Common stem Sc+R+(M)+CuA short, stem Sc+R as long as common stem Sc+R+(M)+CuA; forked at basal 1/3 of forewing length; branch R1 straight, almost reaching the margin, distinctly basad of anteroapical angle, at about 2/3 of forewing length; branch Rs straight at base, curved in apical portion, reaching the margin at anteroapical angle. Stem CuA subparallel to claval suture, directed towards posteroapical angle, not reaching the margin.

Hind wing with costal margin straight, with obtuse apical angles and almost straight apical margin, posterior margin arcuate; margins tuberculate. Costal margin strengthened, with distinct sclerotization at basal 1/3, with a few elongate setae. Stem Sc+R curved at base, sinusoidal to the fork; branch R1 reaching margin basad of anteroapical angle; branch Rs arcuate, reaching margin at anteroapical margin; branch CuA straight, reaching margin at posteroapical angle.

Male genital apparatus elongate, curved dorsad. Claspers with a few setae on external portion.

Etymology

The generic name is derived from "Gapen", the name of the messenger of Baal in Phoenician mythology. Gender: masculine.

Gapenus rhinariatus **sp. n.**
Figs 1–14

Diagnosis

As for genus as it is the only included species and in addition: rhinaria on antennal segments III to X dispersed evenly, distinct, round, with elevated margins; antennal segment III tapering towards base, with a few rhinaria, widest from half of its length and more densely covered with rhinaria in apical half; IIIrd antennal segment with two distinct setae.

Description

Total length of body less than 1 mm, other measurements see Table 1.
Coloration. Forewings and hind wings yellowish, darkened, with brown veins. Body and legs brownish.

Type material

Holotype. Male, specimen No. 1568, Azar collection, provisionally deposited in the collection of Department of Entomology, Muséum national d'Histoire naturelle, Paris, France. The piece of amber has been prepared between two cover slips in Canada balsam medium as described by Azar et al. (2003). The inclusion is compressed and then distorted, which

Table 1. Morphometrics of *Gapenus rhinariatus* gen. sp. n.

Character	Size (mm)
Head with compound eyes width	0.267
Antennal segment I	–
Antennal segment II	–
Antennal segment III	0.123
Antennal segment IV	0.08
Antennal segment V	0.075
Antennal segment VI	0.075
Antennal segment VII	0.051
Antennal segment VIII	0.038
Antennal segment IX	0.028
Antennal segment X	0.032
Forewing length	0.80
Forewing width	0.41
Hind wing length	0.67
Hind wing width	0.31
Fore tibia length	0.21 (right)
Fore basitarsomere length	0.071 (right)
Fore apical tarsomere length	0.052 (right)
Fore claw length	0.018 (right)
Mid tibia length	0.24 (right), 0.25 (left)
Mid basitarsomere length	0.069 (right), 0.07 (left)
Mid apical tarsomere length	0.045 (right), 0.050 (left)
Mid claw	0.013
Hind tibia length	0.30 (right), 0.26 (left)
Hind basitarsomere length	0.079 (right), 0.080 (left)
Hind apical tarsomere length	0.051 (right), 0.049 (left)
Hind claw	0.01 (right), 0.014 (left)

disable the recognition of thoracal structures. It is also strongly curved dorsally with male genital apparatus nearly perpendicular to longitudinal axis of body.

Age and occurrence

Lower Cretaceous, Hammana/Mdeirij, Central Lebanon.

Etymology

Specific epithet refers to the structure of antennal segments with numerous, distinct rhinaria.

Discussion

The newly established genus *Gapenus* gen. n. is placed in Aleurodicinae based on the characters as stem Sc+R forked, with branch R1 preserved and structure of the antennal segments with numerous rhinaria as e.g. in *Palaealeurodicus* Martin, 2008. It is unfortunate, that the other crucial characters are not available for examination. It is interesting that hind wing structure and venation resemble these found in the members of subfamily Udamoselinae. This could support the opinion presented by Martin (2007) that Udamoselinae and Aleurodicinae should be treated as synonyms.

This new genus is for the moment the oldest record of this subfamily, the other one, described as '*Aleurodicus*' *burmiticus* Cockerell, 1919 comes from the fossil resin of younger age, i.e., from the Burmese amber aged also Lower Cretaceous, but Cenomanian, about 90 Ma (Grimaldi et al. 2002, Grimaldi & Engel 2005). This finding indicates that among representatives of Aleyrodidae found as inclusions in the Lower Cretaceous Lebanese amber we have three of four subfamilies represented, i.e., Aleyrodidae, Aleurodicinae and *fide* Shcherbakov (2000)—Udamoselinae.

The newly described extinct genus *Gapenus* gen. n. also sheds new light on morphological disparity and taxonomic diversity of Aleyrodidae from the Lower Cretaceous Lebanese amber forest. The peculiar form of antennal segments, with distinct and numerous rhinaria is an extraordinary feature of the *Gapenus rhinariatus* gen. sp. n.; such developed rhinaria on some segments are known in a few recent species, e.g., *Palaealeurodicus holmesii* (Maskell, 1896) (Martin 2008). Until now little attention has been given to the modifications that have taken place in the structure of whiteflies antennae, but rather to their ultrastructure (Mellor & Anderson

1995a,b; Zheng et al. 2010). Analysing the evolutionary transition of the group it must be assumed that at different times or simultaneously the various processes took place, i.e. reduction of the number of antennal segments, differentiation of the segments, differentiation and transformation of the rhinaria and transformation and differentiation of the setation. The finding of fossil species with particular structure of the antennae as *Gapenus rhinarius* gen. sp. n. described above, gives the new insight into the evolutionary processes of the Aleyrodidae.

Due to the preservation of the inclusion a very little could be said about the other morphological structures of this fossil taxon, however we hope, that with more specimens available for study, the knowledge on diversity and disparity of fossil Aleyrodidae will hopefully increase. We hope also that this encourages the workers and students of recent Aleyrodidae to put more attention into the diversity and the disparity of modern adult whiteflies.

Acknowledgements

We would like to thank Professor Dany Azar, Lebanese University in Beirut for giving us the privilege of studying specimens from Lebanese amber. We wish also to thank Mrs. Marzena Zmarzły for the pencil drawings of the specimen. We would like to thank the Reviewers for carefully reading the manuscript and the helpful comments. This paper is a contribution to project "Insects in the Lower Cretaceous Lebanese amber: taxonomy, palaeobiodiversity and evolution", under cooperative agreements between the Lebanese University (Beirut); MIZ (Museum and Institute of Zoology), Polish Academy of Sciences, Warsaw; University of Gdansk; and the Polish Academy of Sciences, Krakow project "Origin and evolution of biodiversity of Europe and the Middle East" and partly results from the research grant from the Ministry of Science and Higher Education of Poland No. N N303 2979 37, awarded to J.S.

References

Antoine, P.-O., De Franceschi, D. Flynn, J.J., Nel, A., Baby, P., Benammi M., Calderón, N., Espurt, N., Goswami, A. & Salas-Gismondi, R. (2006) Amber from western Amazonia reveals Neotropical diversity during the middle Miocene. *Proceedings of the National Academy of Sciences of the United States of America* **103**: 13595–13600.

Azar, D. (2000) *Les ambres mésozoïques du Liban*. Thèse de Doctorat de l'Université Paris XI, France, 164 pp. + 144 pp., annexes.

Azar, D., Perrichot, V., Néraudeau, D. & Nel, A. (2003) New psychodid flies from the Cretaceous ambers of Lebanon and France, with a discussion about *Eophlebotomus connectens* Cockerell, 1920 (Diptera, Psychodidae). *Annals of the Entomological Society of America* **96**: 117–127.

Cockerell, T.D.A. (1919) Insects in Burmese amber. *The Entomologist* **52**: 241–243.

Drohojowska, J. & Szwedo, J. (2011a) A new whitefly from Lower Cretaceous Lebanese amber (Hemiptera: Sternorrhyncha: Aleyrodidae). *Insect Systematics & Evolution* **42**: 179–196.

Drohojowska, J. & Szwedo, J. (2011b) New Aleyrodidae (Hemiptera: Sternorrhyncha: Aleyrodomorpha) from the Eocene Baltic amber. *Polish Journal of Entomology* **80**: 659–677.

Forero, D. (2008) The systematics of the Hemiptera. *Revista Columbiana de Entomologia* **34**: 1–21.

Grimaldi, D. & Engel, M.S. (2005) *Evolution of the insects*. Cambridge University Press, Cambridge, xv + 755 pp.

Grimaldi, D., Engel, M.S. & Nascimbene, P.C. (2002) Fossiliferous Cretaceous amber from Myanmar (Burma): Its rediscovery, biotic diversity, and paleontological significance. *American Museum Novitates* **3361**: 1–72.

Gullan, P.J. & Martin, J.H. (2003) Sternorrhyncha (Psylloids, Whiteflies, Aphids and Scale Insects). In: Cardé, R.T. & Resh, V.H. (Eds) *Encyclopedia of Insects*. Academic Press, Amsterdam, pp. 1079–1089.

Hennig, W. (1981) *Insect Phylogeny*. (Translated and edited by A.C. Pont). Wiley, Chichester, xii+514 pp.

Koteja, J. & Azar, D. (2008) Scale insects from Lower Cretaceous amber of Lebanon (Hemiptera: Sternorrhyncha: Coccinea). *Alavesia* **2**: 133–167.

Martin, J.H. (2007) Giant whiteflies (Sternorrhyncha, Aleyrodidae): a discussion of their taxonomic and evolutionary significance, with the description of a new species of *Udamoselis* Enderlein from Ecuador. *Tijdschrift voor Entomologie* **150**: 13–29.

Martin, J.H. (2008) A revision of *Aleurodicus* Douglas (Sternorrhyncha, Aleyrodidae), with two new genera proposed for palaeotropical natives and an identification guide to world genera of Aleurodicinae. *Zootaxa* **1835**: 1–100.

Mellor, H.E. & Anderson, M. (1995a) Antennal sensilla of whitefies: *Trialeurodes vaporariorum* (Westwood), the glasshouse whitefly, and *Aleyrodes proletella* (Linnaeus), the cabbage whitefly, and *Bemisia tabaci* (Gennadius), the tobacco whitefly (Homoptera: Aleyrodidae). Part 1: External morphology. *International Journal of Insect Morphology and Embryology* **24**: 133–143.

Mellor, H.E. & Anderson, M. (1995b) Antennal sensilla of whitefies: *Trialeurodes vaporariorum* (Westwood), the glasshouse whitefly, and *Aleyrodes proletella* (Linnaeus), the cabbage whitefly, and *Bemisia tabaci* (Gennadius), the tobacco whitefly (Homoptera: Aleyrodidae). Part 2: Ultrastructure. *International Journal of Insect Morphology and Embryology* **24**: 145–160.

Mound, L.A. & Halsey, S.H. (1978) *Whitefly of the World: a systematic catalogue of the Aleyrodidae (Homoptera) with host plant and natural enemy data*. British Museum (Natural History) and Wiley, Chichester, x + 329 pp.

Poinar Jr., G.O. (1992) *Life in amber*. Stanford University Press, Stanford, CA, xiii + 273 pp.

Poinar Jr., G.O. & Milki, R. (2001) *Lebanese amber: the oldest insect ecosystem in fossilized resin*. Oregon State University Press. Corvallis, OR, 96 pp.

Rietschel, S. (1983) *Aleurochiton petri* n. sp., eine Mottenschildlaus (Homoptera, Aleyrodina) aus dem Pliozän von Neu-Isenburg, Hessen. *Carolinea* **41**: 97–100.

Schlee, D. (1969a) Sperma-übertragung in ihrer Bedeutung für das phylogenetische System der Sternorrhyncha. Phylogenetische Studien an Hemiptera. I. Psylliformes Psyllina und Aleyrodina) als monophyletische Gruppe. *Zeitschrift für Morphologie der Tiere* **64**: 95–138.

Schlee, D. (1969b) Die Verwantschaftsbeziehungen innerhalb der Sternorrhyncha auf Grund synapomorphe Merkmale. Phylogenetische Studien an Hemiptera. II. Aphidiformes (Aphidina–Coccina) als monophyletische Gruppe. *Stuttgarter Beiträge zur Naturkunde* **199**: 1–19.

Schlee, D. (1969c) Bau und Funktion des Aedeagus bei Psyllina und deren Bedeutung für systematische und phylogenetische Untersuchungen (Insecta, Hemiptera). Phylogenetische Studien an Hemiptera. III. Entkräftung eines argument gegen die Monophylie der Sternorrhyncha. *Zeitschrift für Morphologie der Tiere* **64**: 139–150.

Schlee, D. (1970) Verwandtschaftsforschung an fossilien und rezenten Aleyrodina (Insecta, Hemiptera). *Stuttgarter Beiträge zu Naturkunde* **213**: 1–72.

Shcherbakov, D.E. (1990) Extinct four–winged ancestors of scale insects (Homoptera: Sternorrhyncha). In: *Proceedings of the Sixth International Symposium of scale insect Studies*, Part II, Cracow, August 6–12, 1990. Agricultural University Press, Kraków, pp. 23–29.

Shcherbakov, D.E. (2000) The most primitive whiteflies (Hemiptera; Aleyrodidae; Bernaeinae subfam. nov.) from the Mesozoic of Asia and Burmese amber, with an overview of Burmese amber hemipterans. *Bulletin of the Natural History Museum, Geology Series* **56**: 29–37.

Shcherbakov, D.E. (2005) Fossils *versus* Molecules and Cladistics: Controversies over the Hemiptera Phylogeny. In: *Abstracts of Talks and Posters of the 12th International Auchenorrhyncha Congress and concurrent 6th International Workshop on Leafhoppers and Planthoppers of Economic Significance*, University of California, Berkeley, August 7–12, 2005. Opening Keynote Talks: I-1–I-3, available online at http://ag.udel.edu/delpha/4054.pdf (accessed 25 September 2012).

Shcherbakov, D.E. (2007) Extinct four-winged precoccids and the ancestry of scale insects and aphids (Hemiptera). *Russian Entomological Journal* **16**: 47–62.

Shcherbakov, D.E. & Popov, Yu.A. (2002) Superorder Cimicidea Laicharting, 1781 Order Hemiptera Linné, 1758. The Bugs, Cicadas, Plantlice, Scale Insects, etc. (= Cimicida Laicharting, 1781, = Homoptera Leach, 1815 + Heteroptera Latreille, 1810). In: Rasnitsyn, A.P. & Quicke, D.L.J. (Eds) *History of Insects*. Kluwer, Dordrecht, pp. 143–157.

Schmidt, A.R., Perrichot, V., Svojtka, M., Anderson, K.B., Belete, K.H., Bussert, R., Dörfelt H., Jancke, S., Mohr, B., Mohrmann, E., Nascimbene P.C., Nel, A., Nel, P., Ragazzi, E., Roghi, G., Saupe, E.E., Schmidt, K., Schneider, H., Selden, P.A. & Vávra, N. (2010) Cretaceous African life captured in amber. *Proceedings of the National Academy of Sciences of the United States of America* **107**: 7329–7334.

Zheng, L., Wu, W. & Fu, Y. (2010) Sensilla on antenna of adult spiralling whitefly *Aleurodicus dispersus*. *Chinese Bulletin of Entomology* **3**: 525–528.

A new beaded lacewing from a new Lower Cretaceous amber outcrop in Lebanon (Neuroptera: Berothidae)

Dany Azar[a] and André Nel[b]

[a]Lebanese University, Faculty of Sciences II, Department of Natural Sciences, P.O. Box 26110217, Fanar – Matn, Lebanon
E-mail: azar@mnhn.fr
[b]CNRS UMR 7205, Muséum National d'Histoire Naturelle, CP 50, Entomologie, 45 rue Buffon, F-75005 Paris, France
E-mail: anel@mnhn.fr

Abstract

Sibelliberotha rihanensis gen. et sp. n., a new berothid Neuroptera, from Rihan (South Lebanon), a new Lower Cretaceous amber outcrop, is characterized, described, illustrated and its phylogenetic position is discussed. This new fossil taxon possesses several plesiomorphic features that place it very basally within the available phylogeny of modern Berothidae, but close to the modern clade (Nyrminae + Cyrenoberothinae). It enriches our knowledge of the palaeodiversity of this peculiar neuropteran family.

Keywords

Berothidae, phylogeny, gen. et sp. n., Lower Cretaceous, Lebanon, new amber outcrop.

Introduction

The neuropteran family Berothidae (or beaded lacewings) is relatively a small family comprising about 100 living described species in 22 genera, if we do not count the subfamily Rhachiberothinae (Aspöck & Nemeschkal 1998; Archibald & Makarkin 2004). The status of the latter subfamily remains unclear, and is considered as a subfamily of Berothidae (Tjeder 1959; MacLeod & Adams 1967; Grimaldi 2000), or a subfamily of the Mantispidae (Willmann 1990; 1994), or even as a separate family (Aspöck & Mansell 1994; Aspöck & Aspöck 1997). Very recently Makarkin & Kupryjanowicz (2010) reconsidered the position of the rhachiberothids, reconsidering this clade as a subfamily of the Berothidae and also transferred nearly all the fossil rhachiberothids into the berothid subfamily Paraberothinae Nel et al. 2005. For convenience, we follow their classification even if a new phylogenetic analysis of the whole group would be welcome, as the only available one is that of Aspöck & Nemeschkal (1998).

Fossil register of Berothidae is relatively rich with 35 fossil species, described from various localities (see list in Makarkin et al. (2011), with the additions of *Mesithone* Panfilov, 1980 in Makarkin et al. (2012), and *Epimesoberotha* Jepson et al., 2012 (Jepson et al. 2012)).

Herein we describe a new fossil neuropteran berothid from Rihan, a new Lower Cretaceous amber outcrop from Southern Lebanon, which yielded to date to circa 150 biological inclusions, most of them new and very interesting.

Material and Methods

The amber material is originated from Rihan village, Caza Jezzine (Jezzine Department), Mohafazat Loubnan El-Janoubi (South Lebanon Governorate), Southern Lebanon. This outcrop has been discovered on 6 June 2012 during a field trip undertaken by Mr Lutfallah Azar and Miss Eden Hailou in addition to the first author. The outcrop is situated in the beginning of Rihan village in the left side of the road leading to Jezzine, just after getting out from the Ayshiyyeh village. Amber in this outcrop is found as relatively large pebbles (size variant, like walnut or smaller to fist or bigger) in light grey massive sandstone that contain rare lignite remains (Figs 1–2) assigned to Neocomian age (Dubertret 1950, 1955).

The fossil in amber was carefully prepared in Canada balsam medium, following the method described by Azar et al. (2003), in order to observe as many characters as possible. Thus the 'absences' of structures are accurate, which is different of structures that are 'not visible' but maybe present. The specimen was examined with an Olympus SZX9 stereomicroscope and an Olympus inverted compound microscope CK40, both equipped with camera lucida for line drawings and digital cameras.

Systematic Palaeontology

Order Neuroptera Linnaeus, 1758
Family Berothidae Handlirsch, 1906

Figure 1. Rihan outcrop, the day of its discovery, with Mr Lutfallah Azar and Miss Eden Hailou.

Sibelliberotha gen. n.

Type species

Sibelliberotha rihanensis Azar et Nel sp. n.

Etymology

Named after Mrs Sibelle Maksoud, a friend of first author who helped in assembling amber material, and *Berotha*. Gender feminine.

Diagnosis

Face (part between base of antennae and anterior end of clypeus) long (slightly longer than diameter of eyes); galea longer than stipe; presence of a crossvein in forewing area between CuA and MP; pectinate forewing CuA; in all wings, humeral vein small but present; costal area not strongly expanded; quite distinct dark pterostigmata (especially in hind wing, autapomorphy); crossveins in costal area unforked; apical forks of two

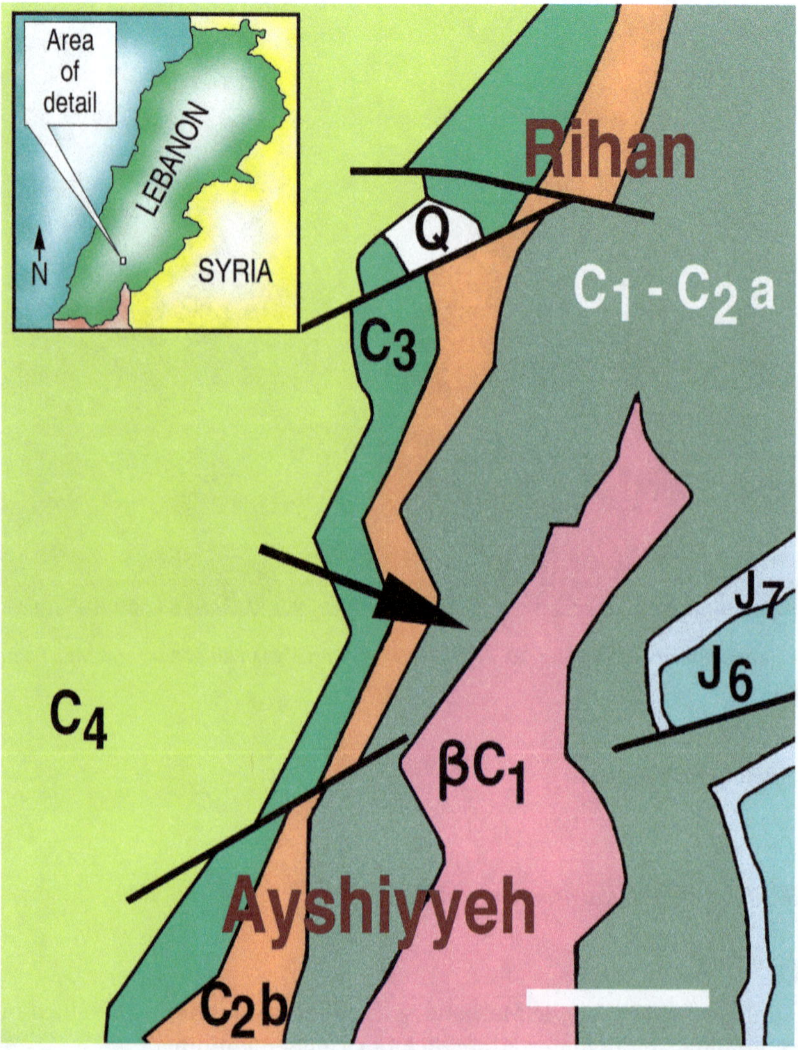

Figure 2. Geological map of Rihan area (adapted from Dubertret 1950). J_6: Kimmeridgian; J_7: Upper Jurassic; C_1–C_2a: Neocomian - Lower Aptian; βC_1: Basaltic Neocomian; C_2b: Upper Aptian; C_3: Albian; C_4: Cenomanian; Q: Quaternary; thickened lines represent faults. The black arrow indicates the amber locality. Scale bar (in white, at right side in the bottom of the map) = 1 km.

Figure 3. Microphotograph of habitus of *Sibelliberotha rihanensis* gen. et sp. n.; female, holotype RIH-12 A, ventral view.

branches of MP very short; very few crossveins in radial area; only two close cells in space between RA and RP; very few branches of radial vein. *Sibelliberotha rihanensis* Azar et Nel sp. n. (Figs 3–11)

Material

Holotype RIH-12A (Fig. 3), Dany Azar coll., provisionally deposited in the "Muséum National d'Histoire Naturelle", Paris, France. The holotype abdomen is parasitised by a mite.

Etymology

After Rihan, type locality of the new amber outcrop.

Diagnosis

As for the genus (*vide supra*).

Locality and horizon

Lower Cretaceous, amber of Rihan (Figs 1–2), Caza Jezzine (Jezzine Department), Mohafazat Loubnan El-Janoubi (South Lebanon Governorate), Southern Lebanon.

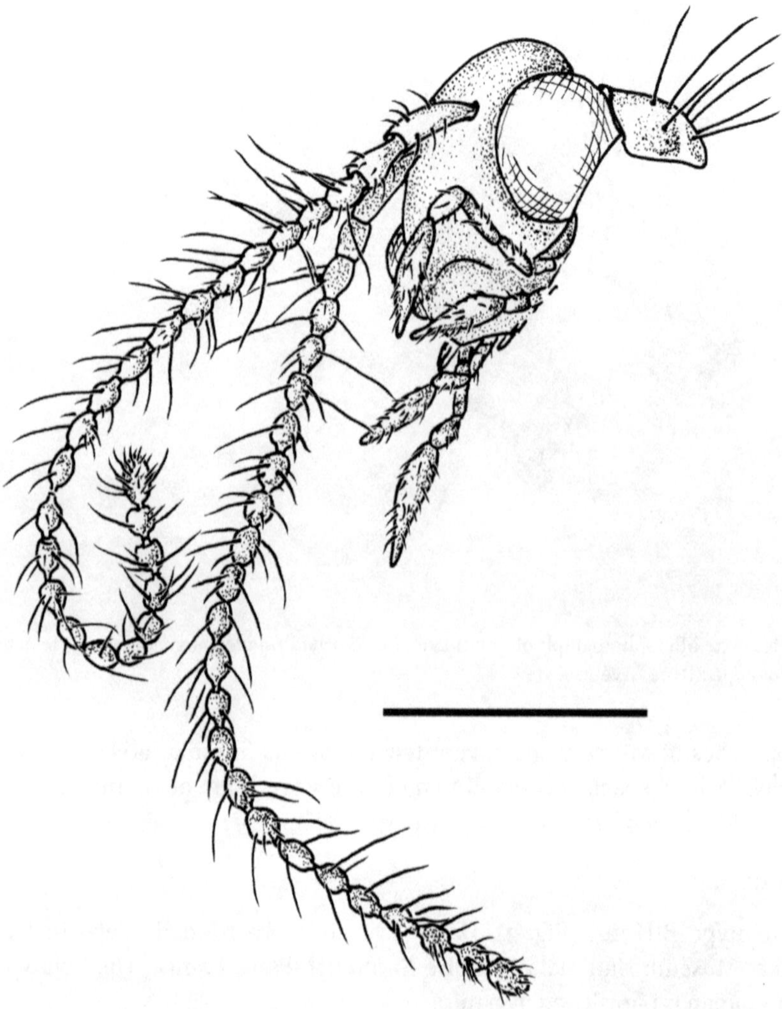

Figure 4. Drawing of head of *Sibelliberotha rihanensis* gen. et sp. n.; female, holotype RIH-12 A, ventral view. Scale bar = 0.3 mm.

Description

Body 2.036 mm long (Fig. 3); head 0.325 mm long (Fig. 4), longer than wide; vertex without any tubercles, no ocelli visible, eyes large, rounded, with small diameter 0.14 mm, great diameter 0.165 mm; mouth parts only partly visible; antennae widely separated, 0.075 mm apart, 1.11 mm long; scape short, 0.1 mm long, 1.5 times as long as pedicel, pedicel broader

Figure 5. Microphotograph of left forewing of *Sibelliberotha rihanensis* gen. et sp. n.; female, holotype RIH-12 A.

Figure 6. Microphotograph of right forewing of *Sibelliberotha rihanensis* gen. et sp. n.; female, holotype RIH-12 A.

than flagellomeres; 22 flagellomeres; flagellomeres setose, all slightly longer than broad; face between base of antennae and anterior end of clypeus slightly longer than diameter of eyes; galea longer than stipe, surpassing level of mandibles but shorter than maxillary palp; maxillary palps 5-segmented, 0.33 mm long; first palpomere 0.025 mm, second palpomere 0.05 mm long, third palpomere 0.065 mm, fourth palpomere 0.05 mm, last palpomere acute at apex, 0.14 mm long; labial palps 3-segmented, 0.2 mm

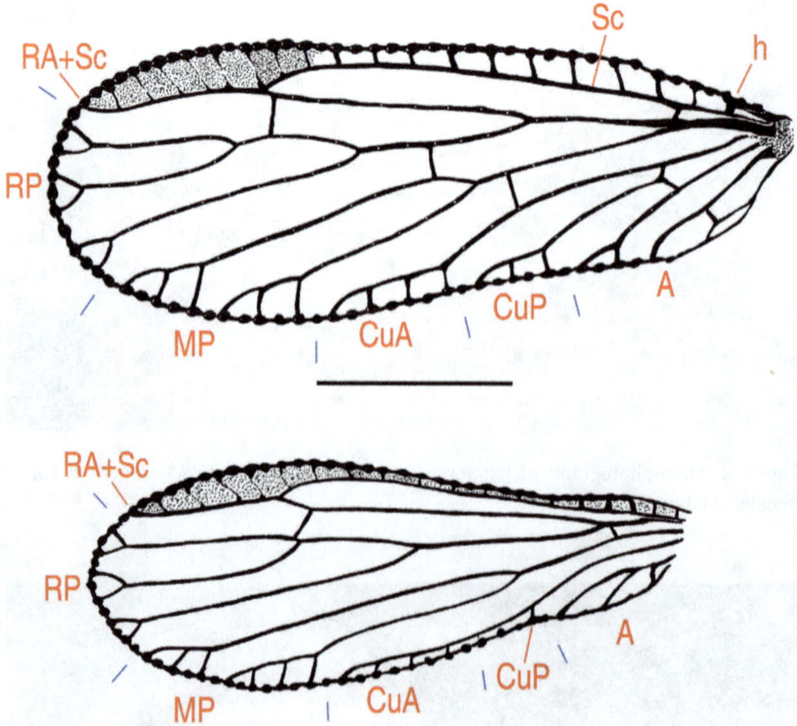

Figure 7. Drawings of wings of *Sibelliberotha rihanensis* gen. et sp. n.; female, holotype RIH-12 A. Scale bar = 0.5 mm.

Figure 8. Microphotograph of right hindwing of *Sibelliberotha rihanensis* gen. et sp. n.; female, holotype RIH-12 A.

Figure 9. Microphotograph of left hindwing of *Sibelliberotha rihanensis* gen. et sp. n.; female, holotype RIH-12 A.

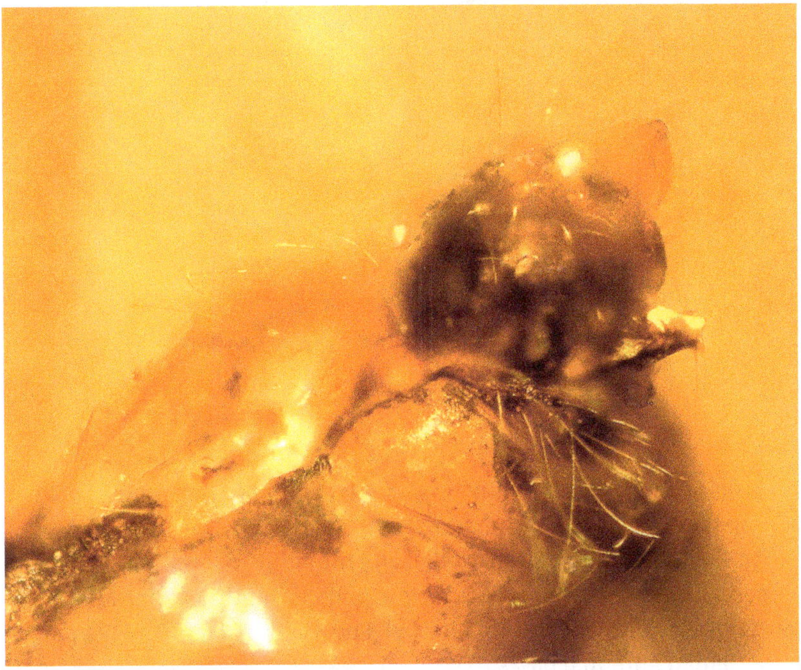

Figure 10. Microphotograph of female genitalia of *Sibelliberotha rihanensis* gen. et sp. n.; holotype RIH-12 A, ventral view.

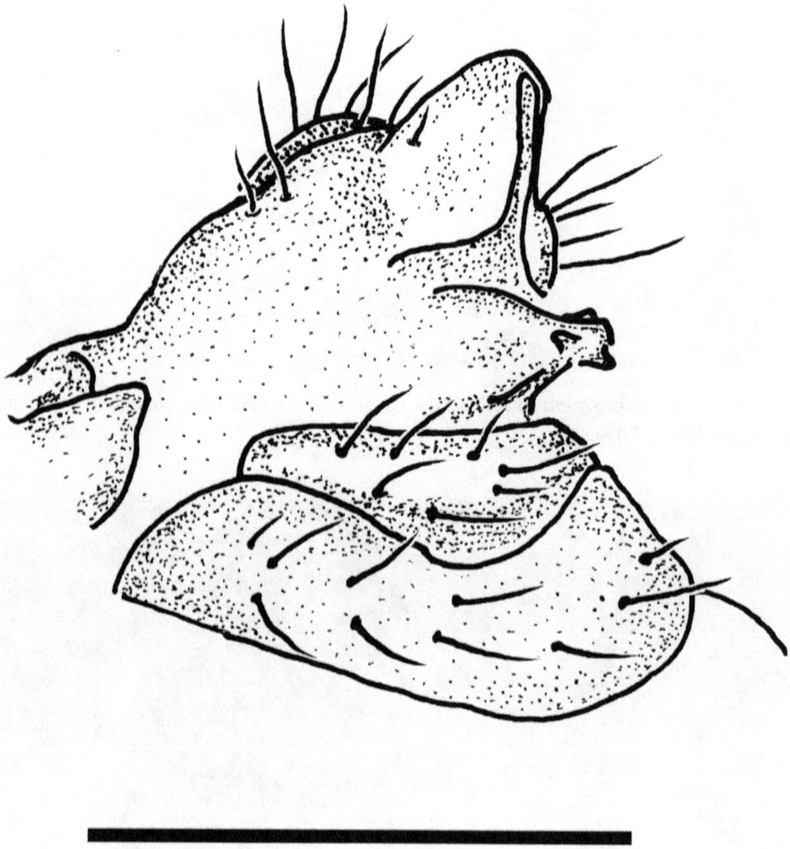

Figure 11. Drawing of female genitalia of *Sibelliberotha rihanensis* gen. et sp. n.; holotype RIH-12 A, ventral view. Scale bar = 0.3 mm.

long; first palpomere 0.05 mm long, second palpomere 0.04 mm, last palpomere acute at apex 0.11 mm long.

Thorax in great part destroyed, 0.454 mm long, only pronotum being preserved, with one transverse furrow and without posterior additional part, not elongate, shorter than broad, and shortening, as long as part of head behind antennal bases.

Fore leg cursorial, devoided of spines; tibia 0.236 mm long, tarsus five segmented; mid leg with numerous small setae on femur and tibia; coxa and trochanter long, femur 0.29 mm long, tibia 0.272 mm long; tarsi five-segmented; hind leg: coax and trochanter long, femur 0.327 mm long, tibia 0.41 mm long; tarsi five-segmented.

Wings densely hairy on veins; fore wing hyaline (Figs 5–7), 1.905 mm long, 0.792 mm wide; with rounded apex; all posterior veins forked near posterior wing margin, with one trichosor between branches along wing margin and at apices of branches; humeral vein simple and small; area between C and Sc with 18 simple cross-veins between h and apex of Sc+RA, narrower than space between Sc and RA, with one visible cross-vein near wing base; Sc and RA distally fused, 0.566 mm basal of wing apex; Sc+RA ending on anterior wing margin 1.811 mm from wing apex; pterostigmal area between Sc+RA and C dark infuscate, with eight small simple cross-veins; base of RP(+MA) 0.679 mm from wing base; RP(+MA) with three posterior branches, one crossvein between RA and RP(+MA) separating two long cells; no cross-veins between the branches of RP(+MA); one cross-vein between MP and RP(+MA); M basally fused with R; fork of M into MP1+2 and MP3+4 both with three very short forks close to posterior wing margin; CuA only with a series of four very short branches along posterior wing margin; CuP only with a series of three very short branches along posterior wing margin; A1 rather long, with two short branches; A2 and A3 simple.

Hind wing hyaline (Figs 7–9), 1.556 mm long, 0.566 mm wide; with rounded apex; all posterior veins forked near posterior wing margin, with only one trichosor between branches and at apices of branches along wing margin; no humeral vein; area between C and Sc infuscate, very narrow, with 18 simple cross-veins; area between Sc and R narrow, with no cross-vein basal of RP(+MA); Sc and RA distally fused, 1.037 mm basal of wing apex; Sc+RA ending on anterior wing margin 0.132 mm from wing apex; pterostigmal area between Sc+RA and C infuscate, with five simple cross-veins; base of RP(+MA) 0.235 mm from wing base; RP(+MA) with three long parallel posterior branches and no series of gradate cross-veins; only one cross-vein between RA and RP(+MA), basal of fusion of RA with Sc; MP separated from R; vein between M and RP(+MA) cross-vein like; fork of MP into MP1+2 and MP3+4 0.424 mm distal of base of M, well distal of base of RP(+MA); MP1+2 with three very short posterior branches close to posterior wing margin while MP3+4 has four; CuA very long, parallel to posterior wing margin, with five very small simple posterior branches; base of CuP not visible; distal part of CuP pectinate, A1, A2, and A3 simple and short.

Abdomen 1.363 mm long, 0.563 mm wide; sternite 7 with posterior margin transverse; fused gonocoxites 8 with one median ventral processus, reduced and not touching tergite 8 laterally, tergite 9 and ectoproct probably separated; female genitalia with pudiculum present (sensu Aspöck & Aspöck 1985) (Figs 10 and 11).

Discussion

The venation of the hind wing of *Sibelliberotha rihanensis* gen. et sp. n. is typical for Berothidae. In particular, the configuration of CuA is characteristic of this family (simple and posteriorly pectinate); although this also occurs in the Nevrorthidae, nevrorthid venation is otherwise dissimilar in the presence of several cross-veins in fore wing area between Sc and R and wing less hairy. Also, the basal crossvein between R and M systems is straight, perpendicular to connecting veins, characteristic of all Berothidae except Rhachiberothinae (and Paraberothinae).

We compare *Sibelliberotha* to the other fossil berothid genera. The rhachiberothine and paraberothine genera can be excluded on the basis of the cursorial forelegs.

Sinosmylites Hong, 1983, *Plesiorobius* Makarkin, 1994, *Mesithone* Panfilov, 1980 (in Berothidae, subfamily Mesithoninae, after Makarkin et al. 2012), and *Epimesoberotha* Jepson et al., 2012 differ from *Sibelliberotha* in the great number of branches of RP, branches of pectinate CuA very long, CuP branched in forewing (Panfilov 1980; Makarkin 1994; Makarkin et al. 2011; Jepson et al. 2012).

Oloberotha Ren & Guo, 1996 and *Banoberotha* Whalley, 1980 differ from *Sibelliberotha* in the forewing forked CuA, three-branched RP (+ MA), costal area strongly expanded in basal third (Whalley 1980; Ren & Guo 1996).

Araripeberotha Martins-Neto & Vulcano, 1990 and *Caririberotha* Martins-Neto & Vulcano, 1990 differ from *Sibelliberotha* in the presence of forked crossveins in costal area, and presence of at least three close cells in space between RA and RP in forewings, plus more radial branches in hind wings (Martins-Neto & Vulcano 1990).

Jersiberotha Engel & Grimaldi, 2008, *Iceloberotha* Engel & Grimaldi, 2008, *Dasyberotha* Engel & Grimaldi, 2008, *Haploberotha* Engel & Grimaldi, 2008, *Jersiberotha* Grimaldi, 2000, *Nascimberotha* Grimaldi. 2000, and *Ethiroberotha* Engel & Grimaldi, 2008 differ from *Sibelliberotha* in the forewing forked CuA, branches of MP with strong secondary forks, and presence of forked crossveins in the area between Sc+RA and Costa (pterostigmal area). Furthermore the three former genera have three close cells in space between RA and RP. *Systenoberotha* Engel & Grimaldi, 2008 has also a forked CuA in forewing. *Telistoberotha* Engel & Grimaldi,

2008 shares with *Sibelliberotha* the pectinate forewing CuA, branches of MP relatively short (even if they are longer in the former), crossveins in costal area all simple, few crossveins in radial area, but it differs in the presence of close cells in area between MP and CuA and of three close cells in space between RA and RP (Grimaldi 2000; Engel & Grimaldi 2008).

Microberotha Archibald & Makarkin, 2004 shows strong similarities in the wing venation with *Sibelliberotha*, with few differences: forks of MP longer with more secondary branches, and a shorter face (Archibald & Makarkin 2004).

Proberotha Krüger, 1923 has forked crossveins in costal area, unlike *Sibelliberotha* (Krüger 1923).

Phylogenetic relationships

We analyse the states of the characters for *Sibelliberotha*, as proposed in the phylogenetic analysis of the recent Berothidae (incl. Rhachiberotinae and Paraberothinae) of Aspöck & Nemeschkal (1998). Nearly all characters with known state are plesiomorphic for the Berothidae, except characters 21, 22, and 50 that are apomorphies of the clade Berothidae nec Rhachiberotinae, the characters 0 and 3 (face long and galea longer than stipe) that are synapomorphies of the clade (*Nyrma*, *Manselliberotha*, *Cyrenoberotha*), and the characters 5, 7, 15, and 49 that are apomorphies subject to several convergencies within the Berothidae, thus less usable for the determination of the affinities of *Sibelliberotha*.

Aspöck & Nemeschkal (1998) considered that the clade (*Nyrma*, *Manselliberotha*, *Cyrenoberotha*) is in a very basal position in the Berothidae, as sister group of Berothidae nec Rhachiberotinae, but they could not solve the trichotomy between these three genera.

Sibelliberotha strongly differs from *Nyrma* (Nyrminae) in the presence of pterostigmata on the four wings, presence of weak humeral vein, all crossveins in costal area simple, and presence of much less crossveins in radial area (Aspöck & Aspöck 1980; Aspöck 1989).

Sibelliberotha also shares with the Cyrenoberothinae (*Manselliberotha*, *Cyrenoberotha*) the presence of a humeral vein (plesiomorphy), but differs by the CuA simple, unforked, crossveins in costal area simple, clearly less branches of RP (+MA) in both fore- and hind wings and less crossveins in radial area (MacLeod & Adams 1967; Aspöck & Aspöck 1988). Aspöck & Nemeschkal (1998) considered their results concerning the phylogeny of the Berothidae as preliminary. We included *Sibelliberotha* in the matrix they proposed and performed a new cladistic analysis using Paup* 4.0b.10, with unordered and equally weighted characters, with the same five taxa they proposed as outgroups (see Appendix for the list of character states in *Sibelliberotha rihanensis* gen. et sp. n., and Figs 12 and 13 for the matrix of characters and strict consensus cladograms). In a first analysis without *Sibelliberotha*, we obtained 398 equally most parsimonious cladograms (Fig. 13A, length=232 steps, C.I. 0.375, R.I. 0.599) using the hsearch option, instead of 64 trees obtained by Aspöck & Nemeschkal (1998: fig. 1). The resulting strict consensus cladogram is not same as the one they obtained, with the main differences concerning the absence of a clade (*Mantispa* + Rhachiberothinae) plus several polytomies in the Berothinae sensus stricto. With the addition of *Sibelliberotha* to this matrix, we obtained 124 equally most parsimonious cladograms (Fig. 13B, length=238 steps, C.I. 0.365, R.I. 0.596), solving the irresolution in part, i.e., we re-obtained the clade (*Mantispa* + Rhachiberothinae) in very basal position together with the outgroup *Nallachius*, plus a 'berothid'clade sensu stricto that comprises in basal positions the taxa *Sibelliberotha*, *Nyrma*, *Manselliberotha*, and *Cyrenoberotha*. This partial analysis confirms the basal position of our fossil in the Berothidae nec Rhachiberotinae. Nevertheless it is obvious that further analyses shall be necessary to solve the berothid phylogeny. At this time, we feel unnecessary to erect a new subfamily for *Sibelliberotha*, even if it has several very particular structures, the presence of hind wing pterostigmata not being the less impressive.

Conclusion

If *Sibelliberotha rihanensis* gen. et sp. n. is in a basal position in the Berothidae nec Rhachiberotinae, it seems nevertheless to be related to recent genera, supporting a great antiquity of these insects. Its inclusion in the existing cladistic analysis based on recent taxa improved its resolution. It would be necessary to include the other fossil berothids in a global phylogenetic analysis, to retrace the evolutionary history of this ancient group.

Acknowledgements

We are thankful to Lutfallah Azar, Eden Hailou, Sibelle, Christine and Marina Maksoud, who helped in assembling the amber material. This paper is a contribution to the team project "Biodiversity: Origin, Structure, Evolution and Geology" awarded to D.A. by the Lebanese University.

Taxon \ Character	1	2	3	4	5	6	7	8	9	10	11	12	13	14	15	16	17	18	19	20	21	22	23	24	25	26	27	28	29
Nallachius	0	1	1	0	0	0	0	0	0	0	0	0	0	0	1	0	1	0	0	0	0	0	0	0	0	0	0	0	0
Trichoscelia	0	0	1	0	0	0	1	0	0	0	1	0	0	0	0	1	1	0	1	0	0	0	0	0	0	0	0	0	0
Mantispa	0	0	0	0	0	2	1	0	1	0	1	0	0	0	1	1	1	0	0	3	0	2	1	2	0	0	0	0	1
Mucroberotha	0	0	0	0	0	0	1	0	0	0	1	0	0	0	0	1	1	0	1	0	0	0	0	2	0	0	0	0	1
Rhachiberotha	0	0	0	0	1	1	1	0	0	0	1	0	0	0	0	0	1	0	1	0	0	0	0	2	0	0	0	0	1
Asedeteva	0	0	0	0	1	1	1	0	1	0	0	2	0	0	1	0	0	0	1	1	0	1	1	0	1	0	1	0	0
Austroberothella	0	0	0	0	0	1	0	1	0	0	0	0	0	0	1	0	1	0	1	0	0	1	1	1	0	0	1	2	0
Berlekrumyia	0	0	0	0	1	1	1	0	1	0	0	2	0	0	0	0	0	0	1	0	0	1	1	2	0	0	1	0	0
Berotha	0	0	0	0	1	1	1	0	0	0	0	2	0	0	1	0	0	0	1	2	0	1	1	1	0	0	1	0	0
Cyrenoberotha	1	0	0	1	0	1	0	0	0	0	0	0	0	0	0	0	1	0	0	0	0	1	1	2	0	0	0	1	1
Isoscelipteron	0	0	0	0	1	1	1	0	1	0	0	2	0	0	1	0	0	0	1	0	0	1	1	1	1	0	1	0	0
Lekrugeria	0	0	0	0	1	1	1	0	1	0	0	1	0	0	0	0	0	0	1	0	0	1	1	1	0	0	1	0	0
Lomamyia	0	0	0	0	1	1	1	0	1	0	0	2	0	0	0	0	1	0	1	0	0	1	1	1	1	0	1	0	0
Manselliberotha	1	0	0	1	0	1	1	0	0	0	0	0	0	0	0	0	1	0	0	0	0	1	1	1	0	0	0	1	1
Naizema	0	1	1	0	0	1	0	1	0	0	0	1	0	0	0	1	1	0	1	0	0	2	1	2	0	0	1	1	0
Nodalla	0	0	0	0	0	1	1	0	1	0	0	0	0	0	1	0	0	0	1	1	0	1	1	1	1	0	1	1	0
Nosybus	0	0	0	0	0	2	0	1	0	0	0	0	0	0	0	1	1	0	1	1	0	1	1	1	0	0	1	1	0
Nyrma	1	0	0	1	0	1	0	0	1	0	0	0	0	1	1	0	1	1	1	0	0	1	0	0	0	0	1	1	1
Podellea	0	0	0	0	2	1	1	0	1	1	0	2	0	0	1	0	0	0	1	0	0	1	1	1	1	0	1	0	1
Protobiella	0	0	0	0	0	1	1	0	0	0	0	0	0	0	0	0	1	0	1	0	1	1	1	1	0	0	1	0	0
Quasispermophorella	0	0	0	0	0	1	1	0	1	0	0	0	0	0	1	1	0	0	1	0	0	1	1	1	1	0	1	0	0
Spermophorella	0	0	0	0	0	1	1	0	0	1	0	0	0	0	1	0	0	0	1	0	0	1	1	1	1	0	1	0	0
Spiroberotha	0	0	0	0	1	1	0	0	1	1	0	1	0	0	0	1	1	0	1	0	0	1	1	1	0	0	1	2	1
Stenobiella	0	0	0	0	1	1	0	0	1	0	0	1	1	0	1	0	0	0	0	2	0	1	1	2	1	1	1	0	0
Trichoberotha	0	1	1	0	0	1	0	1	0	0	0	1	0	0	0	1	0	0	1	0	0	1	1	0	0	0	1	0	0
Trichoma	0	?	?	0	0	1	0	1	0	0	0	1	0	0	0	1	0	0	1	0	0	1	1	0	0	0	1	0	0
Sibelliberotha	1	0	0	1	0	0	0	1	1	0	0	0	0	0	0	1	1	0	0	0	0	1	1	1	0	0	?	?	?

Taxon \ Character	30	31	32	33	34	35	36	37	38	39	40	41	42	43	44	45	46	47	48	49	50	51	52	53	54	55	56	57
Nallachius	1	1	0	1	0	0	0	0	0	0	0	0	0	0	0	0	0	0	0	0	0	0	0	0	0	0	0	1
Trichoscelia	0	1	0	0	0	1	4	4	1	0	1	0	0	1	0	0	0	0	0	2	0	0	1	1	2	1	2	1
Mantispa	0	0	0	0	0	0	0	0	0	0	0	0	0	0	0	0	0	0	0	1	0	0	2	1	2	2	0	0
Mucroberotha	0	0	0	1	0	1	4	4	1	1	0	0	0	1	1	1	1	1	0	2	0	0	0	1	2	0	1	0
Rhachiberotha	0	0	0	1	0	0	0	0	0	0	0	0	0	1	1	1	0	2	0	1	0	0	1	1	2	0	1	0
Asedeteva	1	0	0	0	1	1	2	2	2	0	1	0	0	1	0	0	2	2	0	0	2	2	2	0	1	0	0	0
Austroberothella	0	2	0	0	0	0	1	1	1	0	0	0	0	1	1	1	0	1	0	1	2	1	1	1	1	0	0	0
Berlekrumyia	1	0	0	0	1	1	1	1	1	0	0	0	0	1	0	0	1	1	0	0	2	1	2	0	1	0	0	0
Berotha	1	0	0	1	1	1	2	2	2	0	1	0	0	1	0	0	3	2	1	0	1	2	1	0	1	0	0	0
Cyrenoberotha	0	0	0	0	0	0	0	0	0	0	0	0	0	0	1	0	0	0	0	2	0	1	1	1	1	0	0	0
Isoscelipteron	1	0	0	1	1	1	2	3	2	0	2	1	1	1	0	0	3	2	1	0	1	1	2	0	2	0	0	0
Lekrugeria	0	0	0	1	1	1	2	2	2	0	1	0	0	1	0	0	3	2	0	1	1	1	1	1	1	0	0	0
Lomamyia	0	0	0	1	1	1	1	1	1	0	0	0	0	1	0	0	2	2	1	0	1	1	1	0	1	0	0	0
Manselliberotha	0	0	0	0	0	0	0	0	0	0	0	0	0	0	1	0	0	0	0	2	0	1	1	1	1	0	0	0
Naizema	1	0	0	0	0	0	2	3	1	1	0	0	0	1	0	0	1	1	0	0	0	1	1	0	0	0	0	0
Nodalla	0	0	0	0	1	1	2	2	2	0	1	0	0	1	0	0	2	1	0	0	2	2	2	0	1	0	0	0
Nosybus	1	1	1	0	1	0	2	3	1	2	0	0	0	1	0	0	0	1	0	0	2	2	0	0	2	0	0	0
Nyrma	0	2	0	0	0	0	0	0	0	0	0	0	0	1	0	0	0	0	0	2	0	1	1	0	0	0	0	0
Podellea	0	0	0	2	1	1	2	3	2	0	2	2	0	1	0	0	2	2	0	0	2	2	1	1	2	0	0	0
Protobiella	1	2	0	0	0	0	1	1	1	0	0	0	0	1	1	1	0	0	0	1	2	1	1	1	1	0	0	0
Quasispermophorella	1	0	0	1	1	1	2	3	2	0	2	1	1	1	0	0	3	2	1	0	1	2	2	0	2	0	0	0
Spermophorella	1	0	0	1	1	1	2	3	2	0	2	1	1	1	0	0	3	2	1	0	1	1	2	0	2	0	0	0
Spiroberotha	0	1	0	1	1	0	3	3	1	2	0	0	0	1	1	1	2	2	1	0	2	1	1	0	1	0	0	0
Stenobiella	1	0	0	1	1	1	2	2	2	0	1	0	0	1	0	0	1	0	0	0	2	2	1	0	1	0	0	0
Trichoberotha	1	0	0	1	1	0	1	1	1	0	0	0	1	1	0	0	0	0	0	1	0	1	0	1	0	0	0	0
Trichoma	1	0	0	1	0	1	1	1	0	0	0	1	1	0	0	1	0	0	1	1	1	1	1	0	0	0	0	1
Sibelliberotha	?	?	?	?	?	?	?	?	?	?	?	?	?	0	?	?	?	0	0	1	2	1	?	?	?	0	1&2	0

Figure 12. Matrix of taxa and characters proposed by Aspöck & Nemeschkal (1998), with addition of *Sibelliberotha rihanensis* gen. et sp. n.

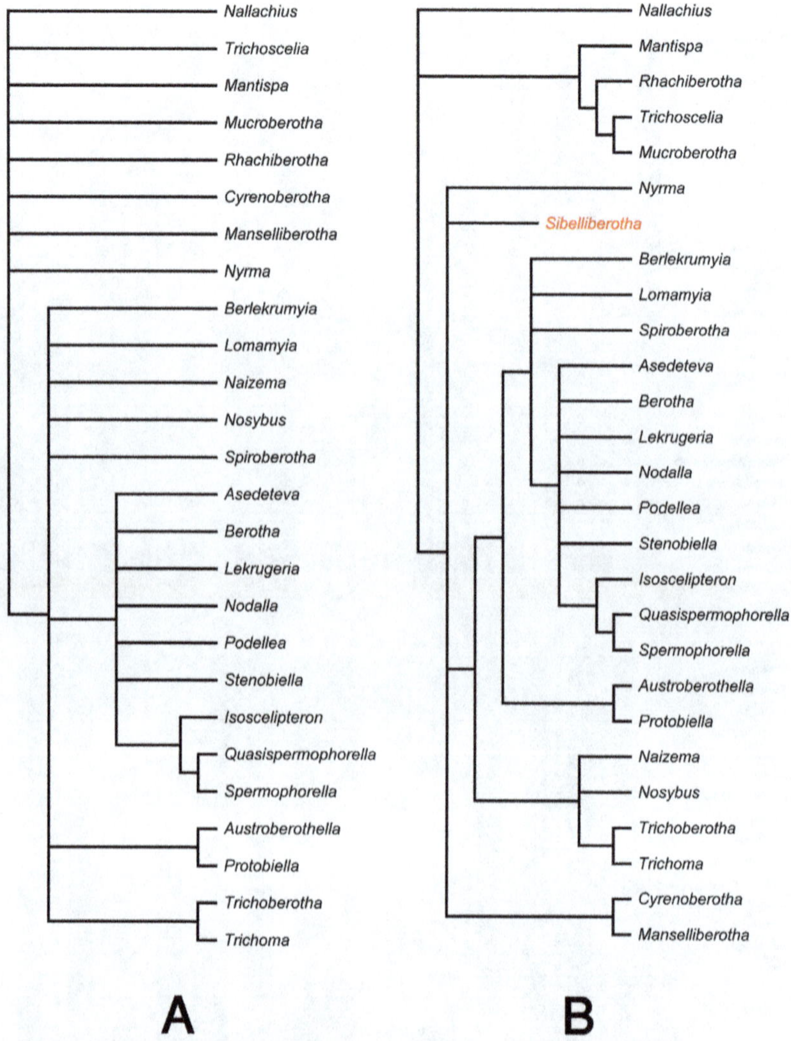

Figure 13. Strict consensus cladograms obtained by cladistic analysis. (A) Without including *Sibelliberotha rihanensis* gen. et sp. n. (B) After addition of *Sibelliberotha rihanensis* gen. et sp. n.

References

Azar, D., Perrichot, V., Néraudeau, D. & Nel, A. (2003) New psychodid flies from the Cretaceous ambers of Lebanon and France, with a discussion about *Eophlebotomus connectens* Cockerell, 1920 (Diptera, Psychodidae). *Annals of the Entomological Society of America* **96**: 117–127.

Archibald, S.B. & Makarkin, V.N. (2004) New genus of minute Berothidae (Neuroptera) from Early Eocene amber of British Columbia. *The Canadian Entomologist* **136**: 61–76.

Aspöck, U. (1989) *Nyrma kervillea* Navás - eine Berothidae! (Neuropteroidea: Neuroptera: Berothidae). *Zeitschrift der Arbeitsgemeinschaft Österreich Entomologen* **41**: 19–24.

Aspöck, U. & Aspöck, H. (1980) *Nyrma kervillea* Navás. Wiederentdeckung einer Systematisch Isolierten Hemerobiiden-Spezies in Kleinasien. *Zeitschrift der Arbeitsgemeinschaft Österreich Entomologen* **31**: 92–96.

Aspöck, U. & Aspöck, H. (1985) Die Berothiden Australiens (und Neuseelands). 2: Die Genera *Trichoma* Tillyard, *Trichoberotha* Handschin, *Protobiella* Tillyard und *Austroberothella* n. gen. (Neuropteroidea: Planipennia: Berothidae). *Zeitschrift der Arbeitsgemeinschaft Österreich Entomologen* **36**: 65–85.

Aspöck, U. & Aspöck, H. (1988) Die Subfamilies Cyrenoberothinae - eine Gondwana-Element? *Manselliberotha neuropterologorum* n. gen. et n. sp. aus S.W.A./Namibia (Neuropteroidea: Planipennia: Berothidae). *Zeitschrift der Arbeitsgemeinschaft Österreich Entomologen* **40**: 1–13.

Aspöck, U. & Aspöck, H. (1997) Studies on new and poorly known Rhachiberothidae (Insecta: Neuroptera) from sub-Saharan Africa. *Annalen des Naturhistorischen Museums in Wien* **99B**: 1–20.

Aspöck, U. & Mansell, MW. (1994) A revision of the family Rhachiberothidae Tjeder, 1959, stat. n. (Neuroptera). *Systematic Entomology* **19**: 181–206.

Aspöck, U. & Nemeschkal, H.L. (1998) A cladistic analysis of the Berothidae (Neuroptera). *Acta Zoologica Fennica* **209**: 45–63.

Azar, D., Perrichot, V., Néraudeau, D. & Nel, A. (2003) New psychodid flies from the Cretaceous ambers of Lebanon and France, with a discussion about *Eophlebotomus connectens* Cockerell, 1920 (Diptera, Psychodidae). *Annals of the Entomological Society of America* **96**: 117–127.

Dubertret, L. (1950) *Carte géologique du Liban au 1/50,000, feuille Marjayoûn, avec notice explicative*. Ministère des Travaux Publics, Beirut.

Dubertret, L. (1955) *Carte géologique du Liban au 1/200,000, avec notice explicative*. Ministère des Travaux Publics, Beirut.

Engel, M.S. & Grimaldi, D.A. (2008) Diverse Neuropterida in Cretaceous amber, with particular reference to the paleofauna of Myanmar (Insecta). *Nova Supplementa Entomologica*, Keltern **20**: 1–86.

Grimaldi, D.A. (2000) A diverse fauna of Neuropterodea in amber from the Cretaceous of New Jersey. In: Grimaldi, D.A. (Ed.). *Studies on fossils in amber, with particular reference to the Cretaceous of New Jersey*. Backhuys Publishers, Leiden, pp. 259–303.

Jepson, J.E., Makarkin, V.N. & Coram, R.A. (2012) Lacewings (Insecta: Neuroptera) from the Lower Cretaceous Purbeck Limestone Group of southern England. *Cretaceous Research* **34**: 31–47.

Krüger, L. (1923) Neuroptera Succinica Baltica. Die im baltischen Bernstein eingeschlossenen Neuropteren des Westpreussichen Provinzial-Museums in Danzig. *Stettiner Entomologische Zeitung* **84**: 68–82.

MacLeod, E.G. & Adams, A. (1967) A review of the taxonomy and morphology of the Berothidae with the description of a new subfamily from Chile. *Psyche* **74**: 237–265.

Makarkin, V.N. (1994) Upper Cretaceous Neuroptera from Russia and Kazakhstan. *Annales de la Société Entomologique de France* (N.S.) **30**: 283–292.

Makarkin, V.N. & Kupryjanowicz, J. (2010) A new mantispid-like species of Rhachiberothinae (Neuroptera: Berothidae) from Baltic amber, with a critical review of the fossil record of the subfamily. *Acta Geologica Sinica* (English Edition) 84: 655–664.

Makarkin, V.N., Yang, Qiang & Ren, D. (2011) Two new species of *Sinosmylites* Hong (Neuroptera, Berothidae) from the Middle Jurassic of China, with notes on Mesoberothidae. *ZooKeys* **130**: 199–215.

Makarkin, V.N., Yang, Q., Peng, Y.-y. & Ren, D. (2012) A comparative overview of the neuropteran assemblage of the Lower Cretaceous Yixian Formation (China), with description of a new genus of Psychopsidae (Insecta: Neuroptera). *Cretaceous Research* 35: 57–68.

Martins-Neto, R.G. & Vulcano, M.A. (1990) Neuropteros (Insecta, Planipennia) da Formação Santana (Cretáceo Inferior), Bacia do Araripe, Nordeste do Brasil. 3. Superfamilia Mantispoidea. *Revista Brasiliera de Entomologia* 34: 619–625.

Panfilov, D.V. (1980) Novye predstaviteli setcharokrylykh (Neuroptera) iz yury Karatau [New representative of Neuroptera from the Jurassic in the Karatau region.]. In: Dolin, V.G., Panfilov, D.V., Ponomarenko, A.G. & Pritykina, L.N. (Eds). *Iskopajemye nasekomyje mezozoja.* [*Fossil insects of the Mesozoic.*] Akademiya nauk Ukrainskoj SSR, Institut Zoologii, Naukova Dumka, Kiev, pp. 82–111. (in Russian)

Ren, D. & Guo, Z.-G. (1996) On the new fossil genera and species of Neuroptera (Insecta) from the Late Jurassic of Northern China. *Acta Zootaxonomica Sinica* 21: 461–479.

Tjeder, B. (1959) Neuroptera - Planipennia. The lace-wings of Southern Africa. 2. Family Berothidae. In: Hanström, B., Brinck, P. & Rudebec, G. (Eds), *South African animal life. Results of the Lund University Expedition in 1950–1951, Volume 6.* Almqvist and Wiksells Boktryckri, Uppsala, pp. 256–314.

Whalley, P.E.S. (1980). Neuroptera (Insecta) in amber from the Lower Cretaceous of Lebanon. *Bulletin of the British Museum (Natural History), Geology* 33: 157–164.

Willmann, R. (1990) The phylogenetic position of the Rhachiberothinae and the basal sister-group relationships within the Mantispidae. *Systematic Entomology* 15: 253–265.

Willmann, R. (1994) Die phylogenetische Position ursprunglicher Mantispidae (Insecta, Planipennia) aus dem Mesozoikum und Alt-Tertiar. *Verhandlungen des Vereins fuer Naturwissenschaftlichen Unterhaltung zu Hamburg* 34: 177–203.

Appendix

List of states of characters present in *Sibelliberotha* (following Aspöck & Nemeschkal 1998)

- Character 0: face (part between base of antennae and anterior end of clypeus) long (slightly longer than diameter of eyes) (state 1) (apomorphy of *Nyrma, Manselliberotha*, and *Cyrenoberotha*).
- Character 1: mouthparts in external position (state 0) (plesiomorphy).
- Character 2: mouthparts not shortening (state 0) (plesiomorphy).
- Character 3: galea longer than stipe (state 1). The galea is long, surpassing the level of the mandibles but shorter than the maxillary palp (apomorphy of *Nyrma, Manselliberotha*, and *Cyrenoberotha*).

— Character 4: antennal scape 1.5 the length of the pedicellus (state 0) (plesiomorphy).
— Character 5: pronotum with one transverse furrow (state 0) "One furrow occurs parallel in *Nallachius, Trichoscelia,* and *Mucroberotha*", so character subject to homoplasy, not very significant.
— Character 6: pronotum not elongate, shorter than broad (state 0) (plesiomorphy).
— Character 7: pronotum shortening (as long as part of head behind antennal bases) (state 1) "shortening of pronotum is apomorphic and interpreted as parallel apomorphy of *Austroberothella,* stem species of TRI, *Nosybus,* and *Naizema*", so character subject to homoplasy, not very significant.
— Character 8: shape of pronotum heavy, with two protuberances visible laterally (state 1).
— Character 9: scales on thorax absent (state 0) (plesiomorphy).
— Character 10: forelegs cursorial (state 0) (plesiomorphy).
— Character 11: wing apex rounded (state 0) (plesiomorphy).
— Character 12: wing distinctly widest in middle third (state 0) (plesiomorphy).
— Character 13: pterostigmal region present (state 0) (plesiomorphy, but as its presence in the hind wing in *Sibelliberotha* is autapomorphic).
— Character 14: forewing recurrent humeral vein present (state 0) (plesiomorphy).
— Character 15: forewing stems of R and M fused before base (state 1) (apomorphy, but subject to homoplasy in the Berothidae).
— Character 16: forewing vein "b", proximal of MP fork absent (state 1) (plesiomorphic for Berothidae).
— Character 17: one forewing sc—r corssvein (state 0) (plesiomorphy).
— Character 18: forewing with one branch of CuA (state 0) (plesiomorphy).
— Character 19: forewing A1 vein with simple dichotomous marginal branching (state 0) (plesiomorphy).
— Character 20: hind wing stems of R and M free (state 0) (plesiomorphy).
— Character 21: hind wing vein "b" crossvein-like (state 1) (autapomorphy Berothidae nec Rachiberothinae).
— Character 22: hind wing base of CuP absent (state 1) (autapomorphy Berothidae nec Rachiberothinae).
— Character 23: hind wing distal part of CuP "creeping" (pectinate) (state 1) (plesiomorphy).

— Character 24: female wing scale absent (state 0) (plesiomorphy).
— Character 25: female abdomen scale (state 0) (plesiomorphy).
— Character 26–41: male characters (unknown for *Sibelliberotha*).
— Character 42: female tergite 9 and ectoproct separated? (state 0) (plesiomorphy).
— Characters 43–45: unknown in *Sibelliberotha.*
— Character 46: female sternite 7 with posterior margin transverse (state 0) (plesiomorphy).
— Character 47: female sternum 7 without pair of sclerite disks (state 0) (plesiomorphy).
— Character 48: female gonocoxites 8 reduced not touching tergite 8 laterally (state 1) (plesiomorphy).
— Character 49: female gonocoxites 8 with one median ventral processus (state 2) (apomorphy but homoplasy in Berothidae).
— Character 50: pudiculum present (state 1) (autapomorphy Berothidae nec Rachiberothinae).
— Characters 51–53: unknown in *Sibelliberotha.*
— Character 54: pronotum without posterior additional part? (state 0) (plesiomorphy).
— Character 55: foretarsus 5-segmented in female (0 or 1) (probably plesiomorphy).
— Character 56 no long ovipositor (state 0) (plesiomorphy).

Nannotanyderus ansorgei sp. n., the first member of the family Tanyderidae from Lebanese amber (Lower Cretaceous)

Wiesław Krzemiński[a], Dany Azar[b,*] and Kornelia Skibińska[a]

[a]Institute of Systematics and Evolution of Animals, Polish Academy of Sciences, Krakow, Poland
[b]Lebanese University, Faculty of Sciences II, Department of Natural Sciences, P.O. Box 26110217, Fanar–Matn, Lebanon
*Corresponding author, e-mail: azar@mnhn.fr

Abstract

A new species, *Nannotanyderus ansorgei*, belonging to Tanyderidae (Diptera, Nematocera), is described and figured from the Lower Cretaceous amber of Lebanon. This is a tiny species, with very particular male genitalia and with wing venation similar to genus *Nannotanyderus krzeminskii* Ansorge, 1994 from Lower Jurassic (Toarcian) of Germany. For the first time a male specimen of the species *Dacochile microsoma* Poinar & Brown, 2004 is illustrated and its genitalia described.

Keywords

Diptera; Tanyderidae; fossil; Lower Cretaceous; Lebanese amber; new taxa; *Nannotanyderus ansorgei*; *Dacochile microsoma*

Introduction

Tanyderidae, or primitive crane flies, is a small, cosmopolitan family of mostly aquatic flies, with about 36 known living species world-wide. The immature stages of the extant species are aquatic to semiaquatic including association with wet sandy soil and the outer layers of submerged rotting logs in streams. Adult males sometimes congregate in large swarms in evenings; spending the daylight hours among the riparian vegetation near the borders of streams. Little is known about the biology of these unusual flies.

Tanyderidae are rather rarely found as fossils. Up till now, representative of six genera have been described, i.e., *Macrochile* Loew, 1850 (two species from Baltic amber (Loew 1850; Alexander 1931; Podenas 1997));

Podemacrochile Krzeminski et Kania, 2013 (Krzeminski et al. 2013) (from Baltic amber); *Nannotanyderus* Ansorge, 1994 (from Lower Jurassic of Germany (Ansorge 1994; Ansorge & Krzemiński 2002)); *Praemacrochile* Kalugina, 1988 (six species from Europe and Asia (Kalugina & Kovalev 1985; Kalugina 1988, 1992; Krzemiński & Ren 2001, Zhang 2004; Lukashevch & Krzemiński 2009)); *Dacochile* Poinar et Brown, 2004 (one species from Cretaceous Burmese amber) and *Protanyderus* Handlirsch, 1909 (which consist of seven extinct species from Jurassic (Kalugina 1988; Lukashevch & Krzemiński 2009)). Genus *Protanyderus* is represented nowadays by eleven extant species from Asia and North America (Alexander 1932; Rose 1963; Exner & Craig 1976; Krzemiński & Judd 1997).

We describe herein a new species that we assign to *Nannotanyderus* (on the basis of wing shape and venation) from the Lower Cretaceous amber of Lebanon. A discovery of a male specimen of *Dacochile microsoma* Poinar et Brown, 2004 from Burmese amber definitively prove the belonging of this species to the Tanyderidae and not to Psychodidae Bruchomyiinae as claimed Woodley (2005).

Material and Methods

The Lebanese amber material described herein is originated from the locality of Jouar Ess-Souss, Bkassine, Caza Jezzine (Jezzine district), Southern Lebanon, Acra Collection. The amber piece containing the inclusion was cut, shaped, and polished, before inclusion between two cover slips in Canada balsam medium as described by Azar et al. (2003). The Burmese amber specimen was generously loaned by Prof. David Grimaldi from the American Museum of Natural History, New York. The specimens were examined with a Nikon SZ10 stereomicroscope, an Olypmus SZX9 steriomicroscope, and a Leitz Laborlux-12 compound microscope, all equipped with digital camera and camera lucida for line drawings.

Systematic Palaeontology

Family Tanyderidae Osten Sacken, 1879
Genus *Nannotanyderus* Ansorge, 1994

Type species

Nannotanyderus krzemińskii Ansorge, 1994, Lower Jurassic (Toarcian).

Nannotanyderus ansorgei **sp. n.** (Figs 1–3)

Holotype

Specimen JG. 385/2B, male; Acra Collection; Lebanese Lower Cretaceous amber from the locality of Jouar Ess-Souss outcrop, Bkassine, Caza Jezzine (Jezzine district), Southern Lebanon; housed provisionally in the "Muséum National d'Histoire Naturelle", Paris, France.

Diagnosis

R_2 very short, shorter than R_3; R_{2+3} 3.5 times longer than R_2, hook like at its basal part, presence of cross-veins m-m and m-cu, presence of cell d, male genitalia well developed with specific elongated gonocoxite basistylus and gonostylus which reaches about 3/4 abdomen length.

Etymology

Named in honour of the German paleoentomologist Dr Jörg Ansorge, who described the most important fossil Tanyderidae and Psychodidae from Lower Jurassic.

Description

Very small specimen, 1.89 mm long (Fig. 1). Head 0.33 mm wide; eyes large, 0.29 mm of diameter; antennae long, delicate; palpus long with last segment 2 times longer than penultimate. Thorax 0.6 mm long; wing hyaline, without stigma (Fig. 2), 1.56 mm long, 0.59 mm wide; h reaching wing margin at 0.29 mm from wing base; Sc short, 0.81 mm long; cross-vein sc-r present; R_1 ending in wing margin at about 1.15 mm from wing base, opposite to fork of R_{2+3} into R_2 and R_3; R_{2+3} 2 times longer than Rs and 2 times longer than R_3; Rs only little longer than R_{4+5} and R_4 more than 3 times longer than R_{4+5}; M_1 the same length like d cell; d cell closed; cross-vein m-cu leaves M_4 after fork of M_{3+4} into M_3 and $M_{4:}$ a_1 cell narrow. The cell between M_4 and Cu on the wing border is four times longer than the one of a_1 cell (Fig. 2). Legs long with two apical tibial spurs. Abdomen, 0.69 mm, with very long male terminalia of inverted type (like most of Psychodidae) (Fig. 3); gonopods very long and narrow, nearly as long as the whole abdomen; gonocoxite well developed, with strong bristles on inner margin; gonostylus very long, nearly the same length like gonocoxite, with hook like and bifurcate terminal part; aedeagal complex rather small, but badly visible (Fig. 3).

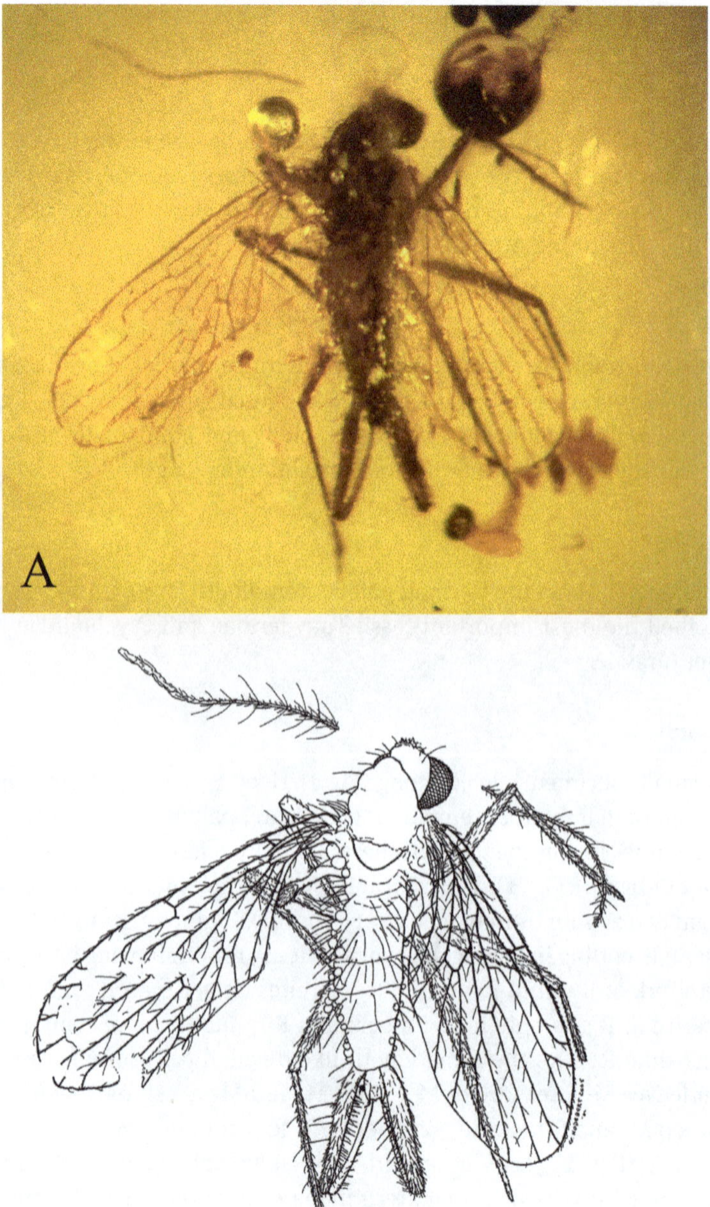

Figure 1. *Nannotanyderus ansorgei* sp. n., holotype (JG. 385/2B), male; habitus. (A) Photograph. (B) Line drawing. Scale bar=1 mm.

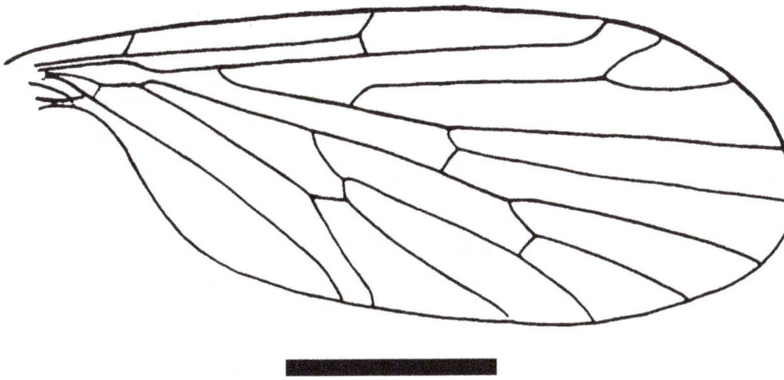

Figure 2. Reconstruction of wing of *Nannotanyderus ansorgei* sp. n., holotype (JG. 385/2B). Scale bar=0.5 mm.

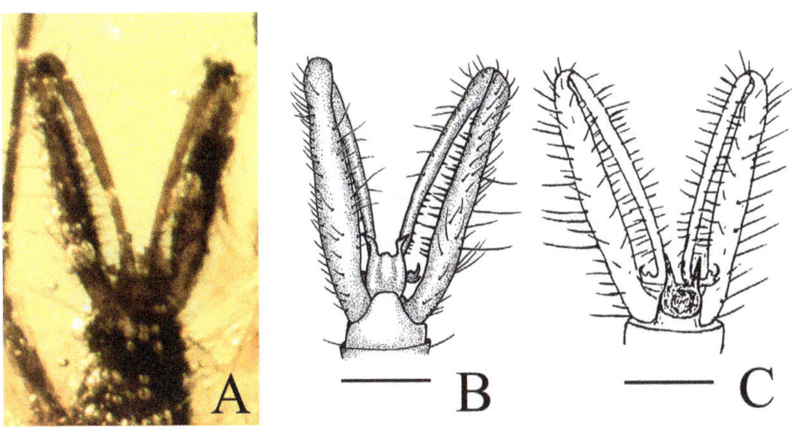

Figure 3. Male genitalia of *Nannotanyderus ansorgei* sp. n., holotype (JG. 385/2B). (A) Photograph. (B) Line drawing of dorsal side. Scale bar=0.2 mm. (C) Line drawing of ventral side. Scale bar=0.2 mm.

Dacochile microsoma Poinar et Brown, 2004 (Figs 4–8)

Material

Specimen Bu 1262, very nicely preserved male (Fig. 4), Burmese late Albian–early Cenomanian amber (Shi et al. 2012), Hukawng Valley, housed in the American Museum of Natural History, New York, NY, USA.

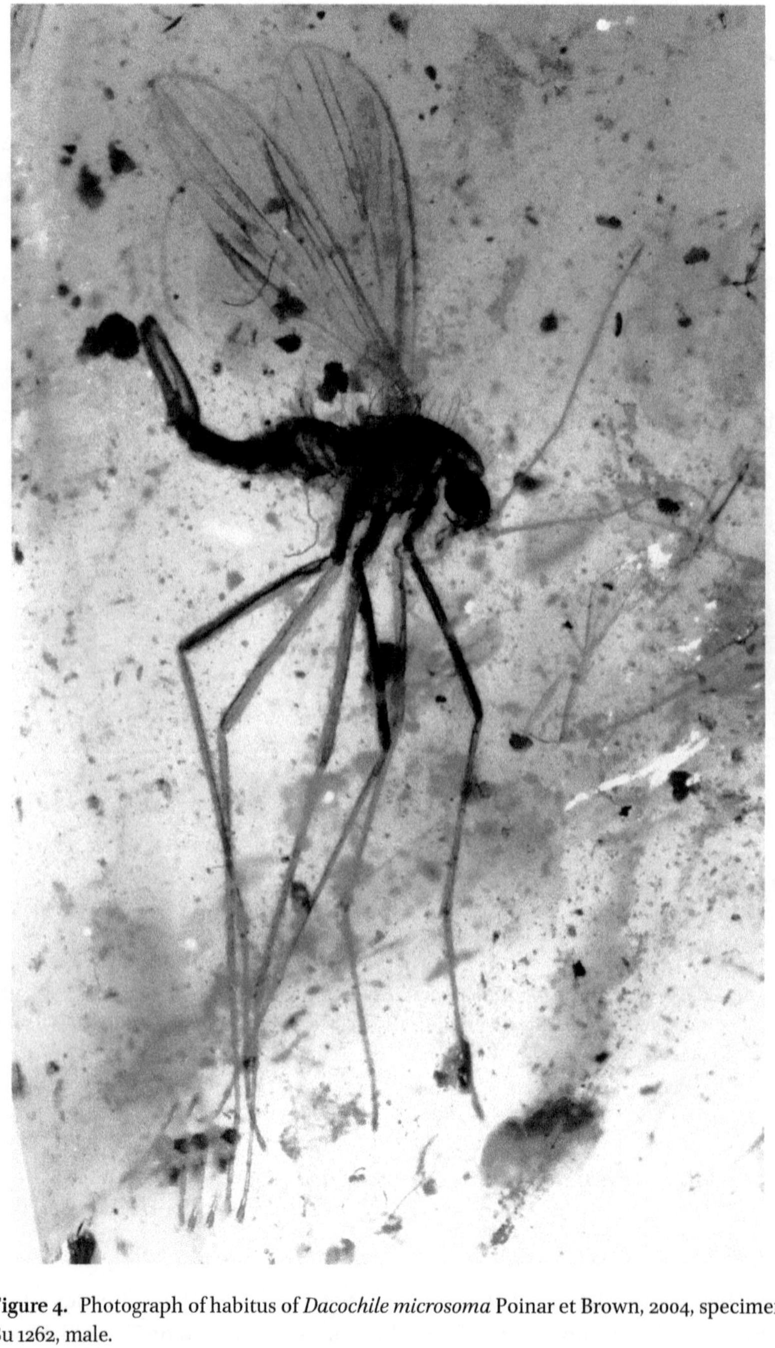

Figure 4. Photograph of habitus of *Dacochile microsoma* Poinar et Brown, 2004, specimen Bu 1262, male.

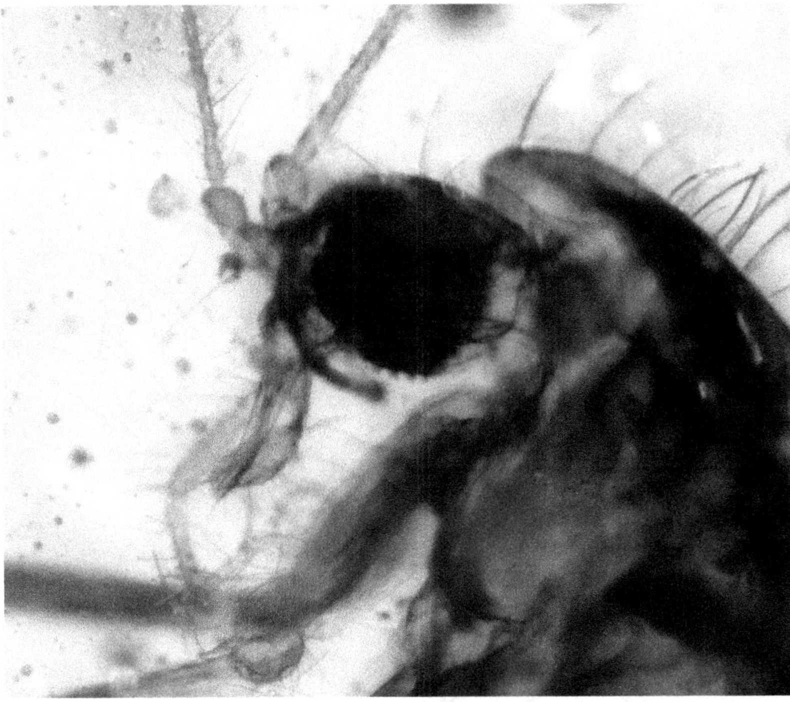

Figure 5. Photograph of head of *Dacochile microsoma* Poinar et Brown, 2004, male, specimen Bu 1262.

This species, belonging to a monospecific genus, was described by Poinar & Brown (2004) based on a single female specimen and attributed to tanyderids. We report here the discovery of a male specimen belonging to *Dacochile microsoma,* with all the morphological features as described by Poinar & Brown (2004, 2006), including head with mouthparts with well-developed mandibles (Figs 5 and 6) (but no serration were discernible); and wing venation arrangement (Fig. 7) and Male terminalia are inverted, well developed (Fig. 8) and present a lot of similarity with *Nannotanyderus ansorgei* sp. n.

Discussion and conclusion

Ansorge (1994) described genus *Nannotanyderus* only based on wing venation (Fig. 9). The Lebanese amber specimen is a well-preserved male member with wings' venation, somehow comparable to Jurassic specimen *Nannotanyderus krzeminskii* Ansorge (1994). *Nannotanyderus ansorgei*

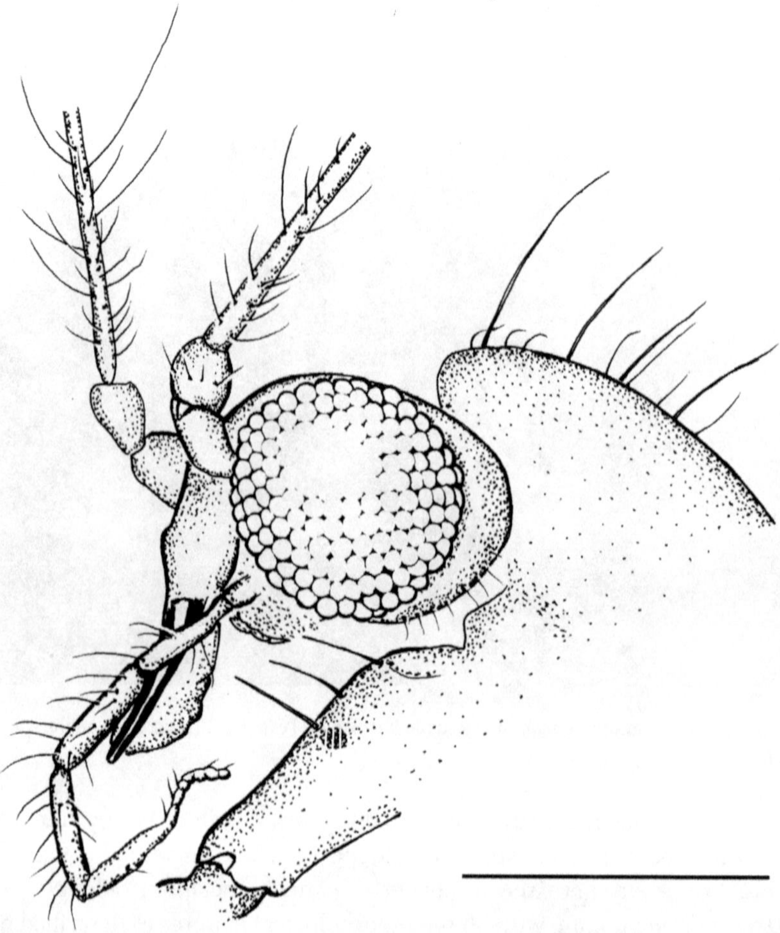

Figure 6. Line drawing of head of *Dacochile microsoma* Poinar et Brown, 2004, male, specimen Bu 1262. Scale bar=0.3 mm.

sp. n. shares some features with psychodids (small size, absence of well-developed anal lobe, inverted genitalia, fifth palpal segment considerably longer than fourth, wings immaculate) and others with tanyderids (arrangement of setae on flagellomeres absence of ascoids, presence of cross-vein m-m, presence of m-cu cross-vein, presence of cell d, presence of tanyderid structure of the male genitalia, spurs on the tibiae, etc.), but based on the cited characters the combination would orient rather to tanyderids. Today a global phylogeny of the psychodomorphan including

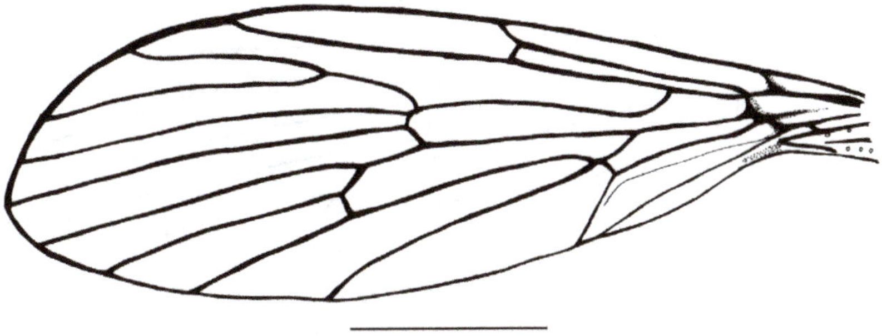

Figure 7. Line drawing of wing of *Dacochile microsoma* Poinar et Brown, 2004, male, specimen Bu 1262. Scale bar=0.5 mm.

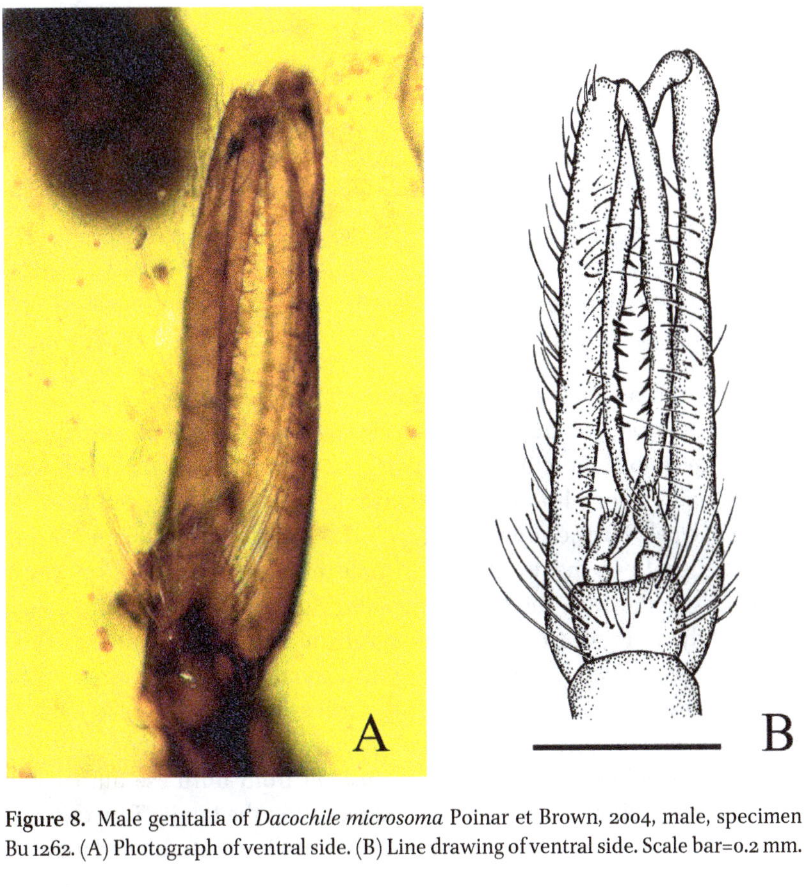

Figure 8. Male genitalia of *Dacochile microsoma* Poinar et Brown, 2004, male, specimen Bu 1262. (A) Photograph of ventral side. (B) Line drawing of ventral side. Scale bar=0.2 mm.

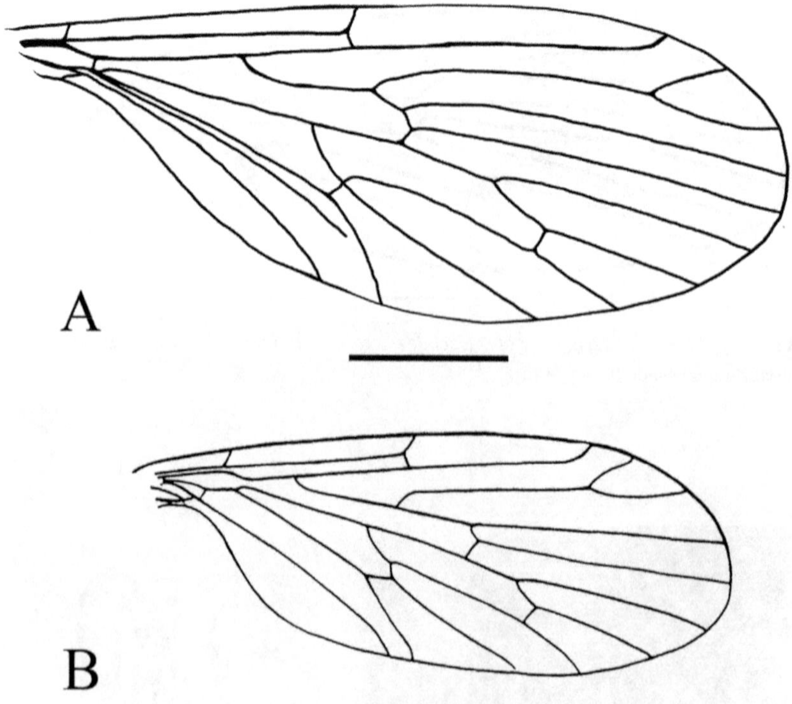

Figure 9. Comparison of wings of *Nannotanyderus krzemińskii* and *Nannotanyderus ansorgei*. (A) Wing of *Nannotanyderus krzemińskii* Ansorge, 1994. Scale bar=0.5 mm. (B) wing of *Nannotanyderus ansorgei* sp. n., holotype (JG. 385/2B). Scale bar=0.5 mm.

recent and fossil taxa is a need in order to define the real phylogenetic value of the used characters. The species *Nannotanyderus ansorgei* sp. n. (Fig. 2) described herein is different from *Nannotanyderus krzemińskii* (Fig. 9) in: Sc is much longer in *N. ansorgei*; and cross-vein r-m is in 2/3 length of d cell. Furthermore very good preservation of the Lebanese specimen amber allows the recognition of male terminalia construction. These are very specific with the long gonapods. Moreover male terminalia are completely different from remaining Tanyderidae, which are short and wide except *Peringueyomyina barnardi* Alexander, 1921, but here too the genitalia are not inverted whereas in our fossil it is inverted as most of Psychodidae. This fossil would be the smallest known tanyderid.

Female specimen of *Dacochile microsoma* from Burmese amber was described by Poinar & Brown 2004 and placed in the family Tanyderidae. Woodley (2005) challenged Poinar & Brown opinion and assigned it rather to the subfamily of Bruchomyiinae within the Psychodidae based on some features present in the latter. After that another female specimen was found and enabled Poinar & Brown to confirm their original attribution for this very polemic and enigmatic species (Poinar & Brown 2006). The discovery of our male *Dacochile microsoma* confirms its alignment with Tanyderidae, even if this particular species shares several features with Bruchomyiinae (Psychodidae) as indicated by Woodley (2005). It is worth to indicate that Bruchomyiinae had been classified within the Tanyderidae by Alexander (1920), before this author adopted definitively the actual classification.

The two fossils subject of this paper, are the smallest known tanyderids. However the discovery of very small Tanyderidae and primitive Psychodidae from the Jurassic and Cretaceous periods (Ansorge 1994; Poinar & Brown 2004), allow us to better understand the close phylogenetic relationship between Tanyderidae and Psychodidae as stated in Krzemiński (1992), Krzemiński & Evenhuis (2000) and Krzemiński & Krzemińska (2003). Recent molecular data of Bertone et al. (2008) bring additional evidence to support the hypothesis of very close relationship between Tanyderidae and Psychodidae. The discovery of male specimen of *Dacochile microsoma* confirms the attribution of this species to the family Tanyderidae based on the existing combination of features.

Although the existence of possibility of confusion in the classification of today tanyderids and psychodids with the discovery of the fossil material, we prefer for the moment to keep the classification as is, and to assign *Nannotanyderus ansorgei, Dacochile microsoma* to the Tanyderidae. Only finding more fossil and recent material would give great clue to understand the real evolutionary history of the tanyderids and psychodids.

Acknowledgements

The authors are thankful to unknown reviewers for their instructive comments on a previous version of this paper. We thank Prof. David Grimaldi for loaning the Burmese amber material and Mr. Gilbert Hodebert for *Nannotanyderus ansorgei* habitus artwork. This paper is a contribution to the project "Insects in the Lower Cretaceous Lebanese amber: taxonomy, palaeobiodiversity and evolution", under cooperative agreements between the Lebanese University (Beirut), MIZ (Museum and Institute of Zoology), Polish Academy of Sciences, Warsaw, University of Gdansk, and the Polish Academy of Sciences, Krakow project "Origin and evolution of biodiversity of Europe and the Middle East". This paper is a contribution to the team project "Biodiversity: Origin, Structure, Evolution and Geology" awarded to D.A. by the Lebanese University.

References

Alexander, C.P. (1920) A new subfamily of tanyderid flies (Diptera). *Annals of the Entomological Society of America* **13**: 402–406 + Plate XXXII.

Alexander, C.P. (1921) A new genus and species of Tanyderidae (*Peringueyomyina barnardi*) in the South African museum (Diptera). *Annals of the South African Museum* **18**: 231–234.

Alexander, C.P. (1931) Crane-flies of the Baltic amber (Diptera). *Bernstein-Forschung* **2**: 1–135.

Alexander, C.P. (1932) The Dipterous family Tanyderidae in Japan (Insecta). *Annotationes Zoologicae Japonenses* **13**: 273–281.

Ansorge, J. (1994) Tanyderidae and Psychodidae (Insecta: Diptera) fom the Lower Jurassic of northeastern Germany. *Paläontologische Zeitschrift* **68**: 199–210.

Ansorge, J. & Krzemiński, W. (2002) Lower Jurassic tanyderids (Diptera: Tanyderidae) from germany. *Studia Dipterologica* **9**: 21–29.

Azar, D., Nel, A., Solignac, M., Paicheler, J.-C. & Bouchet, F. (1999) New genera and species of psychodoid flies from the Lower Cretaceous amber of Lebanon. *Palaeontology* **42**: 1101–1136.

Azar, D., Perrichot, V., Néraudeau, D., & Nel, A. (2003) New psychodid flies from the Cretaceous ambers of Lebanon and France, with a discussion about Eophlebotomus connectens Cockerell, 1920 (Diptera, Psychodidae). *Annals of the Entomological Society of America* **96**: 117–127.

Bertone, M.A., Courtney, G.W. & Wiegmann, B.M. (2008) Phylogenetics and temporal diversification of the earliest true flies (Insecta: Diptera) based on multiple nuclear genes. *Systematic Entomology* **33**: 668–687.

Exner, K. & Craig, D.A. (1976) Larvae of Alberta Tanyderidae (Diptera: Nematocera). *Quaestiones Entomologicae* **12**: 219–237.

Handlirsch, A. (1909) Zur Phylogenie und Flügelmorphologie der Ptychopteriden (Dipteren). *Annalen der Naturhistorischen Hofmuseums, Wien* **23**: 263–272.

Kalugina, N.S. (1988) [Mesozoic Diptera: Psychodomorpha and Tipulomorpha (Tanyderidae, Eoptychopteridae, Limnoniidae).]. In: *Trudy Sovmestnaya Sovetsko-Mongol'skaya Paleontologicheskaya Ehkspeditsiya* [New species of fossil invertebrates of Mongolia.] **33**: 81–89. (in Russian)

Kalugina, N.S. (1992) Psychomorphan Diptera from the Jurassic of Mongolia Altai (Diptera: Tanyderidae, Eoptychopteridae). *Paleontological Journal* **26**: 110–113.

Kalugina, N.S. & Kovalev, V.G. (1985) [Dipterous insects of Jurassic Siberia]. Paleontological Institute, Akademia Nauk, Moscow, 198 pp. (in Russian)

Krzemiński, W. (1992) Triassic and Lower Jurassic stage of Diptera evolution. *Mitteilungen der schweizerischen entomologischen Gesellschaft* **65**: 39–59.

Krzemiński, W. & Evenhuis, N.L. (2000). Review of Diptera palaeontological records. In: Papp, L. & Darvas, B. (Eds). *Manual of Palaearctic Diptera, Volume 1. General and Applied Dipterology*. Science Herald, Budapest, pp. 535–564.

Krzemiński, W. & Judo, D.D. (1997) Chapter 2.14. Family Tanyderidae. In: Papp, L. and Darvas, B. (Eds). *Manual of Palaearctic Diptera, Volume 2*. Science Herald, Budapest, pp. 281–289.

Krzemiński, W. & Krzemińska, E. (2003) Triassic Diptera: descriptions, revisions and phylogenetic relations. *Acta Zoologica Cracoviensia* **46** (Suppl. Fossil Insects): 153–184.

Krzemiński, W. & Ren, D. (2001) *Praemacrochile chinensis* sp. n. from the Middle Jurassic of China (Diptera: Tanyderidae). *Polskie Pismo Entomologiczne* **70**: 127–129.

Krzeminski, W., Krzeminska, E., Kania, I. & Ross, A.J. (2013) New taxa of Tanyderidae (Diptera) from Eocene Baltic amber. *Zootaxa* **3599**: 59–66.

Loew, H. (1850) Dipterologische Beiträge. *Vierter Beitrag. Die Gallmücken. (Progr). K. Friedrich-Wilhelms-Gymnasiums Posen*: 1–40.

Lukashevich, E.D. & Krzemiński, W. (2009) New Jurassic Tanyderidae (Diptera) from Asia with first find of larvae. *Zoosymposia* **3**: 155–172.

Osten-Sacken, C.R. (1879) Die Tanyderina, eine merkwürdige Gruppe der Tipuliden. *Verhandlungen der Kaiserlich-Königlichen zoologisch-botanischen Gesellschaft in Wien* XXIX: 517–522.

Poinar Jr., G.O. & Brown, A.E. (2004) A new genus of primitive crane flies (Diptera: Tanyderidae) in Cretaceous Burmese amber, with a summary of fossil tanyderids. *Proceedings of the Entomological Society of Washington* **106**: 339–345.

Poinar Jr., G.O. & Brown, A.E. (2006) The enigmatic *Dacochile microsoma* Poinar & Brown: Tanyderidae or Bruchomyiinae?. *Zootaxa* **1162**: 19–31.

Shi. G., Grimaldi, D.A., Harlow, G.E., Wang, J., Wang, J., Yang, M., Lei, W., Li, Q. & Li, X. (2012) Age constraint on Burmese amber based on U–Pb dating of zircons. *Cretaceous Research* **37**: 155–163.

Woodley, N.E. (2005) *Dacochile microsoma* Poinar & Brown, not a tanyderid but a bruchomyiine psychodid (Diptera: Psychodidae, Tanyderidae). *Zootaxa* **1012**: 53–60.

Zhang, J.-F. (2004) Nematoceran dipterans from the Jurassic of China (Insecta, Diptera: Limoniidae, Tanyderidae). *Paleontological Journal* **38**: 522–527.

First record of Perforissidae from the Early Cretaceous Lebanese amber (Hemiptera: Fulgoromorpha: Fulgoroidea)

Jacek Szwedo[a,*], Dany Azar[b] and Youssef Nohra[b,c]

[a]Department of Palaeozoology, Museum and Institute of Zoology, Polish Academy of Sciences, Wilcza Street 64, PL00-679 Warszawa, Poland
[b]Lebanese University, Faculty of Sciences II, Department of Biology, P.O. Box 26110217, Fanar–Matn, Lebanon
[c]Géosciences Rennes (UMR CNRS 6118), Université de Rennes 1, Campus de Beaulieu, Building 15, Avenue du General Leclerc 263, 35042 Rennes cedex, France
*Corresponding author, e-mail: szwedo@miiz.waw.pl

Abstract
A new genus, *Aafrita* gen. n., with a new species, *Aafrita biladalshama* sp. n., from the Early Cretaceous Lebanese amber is described. It is the first record of the extinct family Perforissidae (Hemiptera: Fulgoroidea) from the fossil resins of Gondwanaland. Imagines (females), nymphs and exuvia were found in three outcrops in Northern, Central and Southern Lebanon. The morphological peculiarities of the fossils are briefly discussed. FT-IR spectra of the fossil resins in which the fossils were found are presented and discussed.

Keywords
Aafrita gen. n.; *Aafrita biladalshama* sp. n.; new genus; new species; Lebanese amber; Early Cretaceous; taxonomy; morphology; fossil resins; FT-IR analysis

Introduction

Planthoppers (Hemiptera: Fulgoromorpha) enter the fossil record since the Permian (Shcherbakov & Popov 2002; Szwedo et al. 2004). This hemipteran infraorder includes Permian Coleoscytoidea Martynov, 1935, (with Coleoscytidae Martynov, 1935), Permian and Triassic Surijokocixioidea Shcherbakov, 2000 (with Surijokocixiidae Shcherbakov, 2000) and Fulgoroidea Latreille, 1807, known since the Jurassic. The latter comprises over 20 families, extinct and extant (Bourgoin 2013). Extinct families of Fulgoroidea are Jurassic Qiyangiricaniidae Szwedo, Wang et Zhang, 2011 and paraphyletic 'Fulgoridiidae' Handlirsch, 1906 (Bourgoin & Szwedo 2008), Cretaceous ones, Lalacidae Hamilton, 1990, Neazoniidae Szwedo, 2007, Perforissidae Shcherbakov, 2007 and Mimarachnidae Shcherbakov, 2007, and a single extinct family from the Palaeogene, Weiwoboidae Lin, Szwedo, Huang et Stroiński, 2010 (Hamilton 1990; Lin et al. 2010; Shcherbakov 2007a,b, Szwedo 2007, 2008, 2009; Bourgoin & Szwedo 2008; Szwedo et al. 2011). The extant Fulgoroidea families are Acanaloniidae Amyot et Serville, 1843; Achilidae Stål, 1866 + Achilixiidae Muir, 1923; Caliscelidae Amyot et Serville, 1843; Cixiidae Spinola, 1839; Delphacidae Leach, 1815; Derbidae Spinola, 1839; Dictyopharidae Spinola, 1839; Eurybrachidae Stål, 1862 + Gengidae Fennah, 1949; Flatidae Spinola, 1839 + Hypochthonellidae China et Fennah, 1952; Fulgoridae Latreille, 1807; Issidae Spinola, 1839; Kinnaridae Muir, 1925 + Meenoplidae Fieber, 1872; Lophopidae Stål, 1866; Nogodinidae Melichar, 1898; Ricaniidae Amyot et Serville, 1843; Tettigometridae Germar, 1821; and Tropiduchidae Stål, 1866 (O'Brien 2002; Bourgoin 2013). The monophyly, relationships, range and content of several families (both extant and extinct) are under discussion and there is no consensus on phylogenetic relationships within this superfamily. Two informal groups, so-called 'Cixiidae-like' planthoppers and 'higher Fulgoroidea' are sometimes indicated, but there are no clear definitions of these groups (Bourgoin & Szwedo 2008, Lin et al. 2010).

The family Perforissidae is one of the groups known only from the Cretaceous (Shcherbakov 2007a, Peñalver & Szwedo 2010) and it is placed among the 'Cixiidae-like' planthoppers (Szwedo 2009). It is a peculiar group, which presents a combination of particular morphological characters not to be found among either extinct or extant Fulgoroidea.

Material and Methods

The specimens described here originate from the three outcrops located in Northern, Central and Southern Lebanon (Figs 1–3). The amber outcrops are Neocomian in age; amber is associated with lignite and plant debris in a layer of dark clay. Few inclusions are presently reported from this fossil site (Azar et al. 2010; Azar 2012).

All three spectra of the amber pieces with inclusions described below present similar profiles (Fig. 4), although they are provided from different geographical locations (North, Central and South Lebanon). This similarity could be due to a probable same botanical origin for the resin, exuded and deposited in the same environmental conditions. A slight but yet

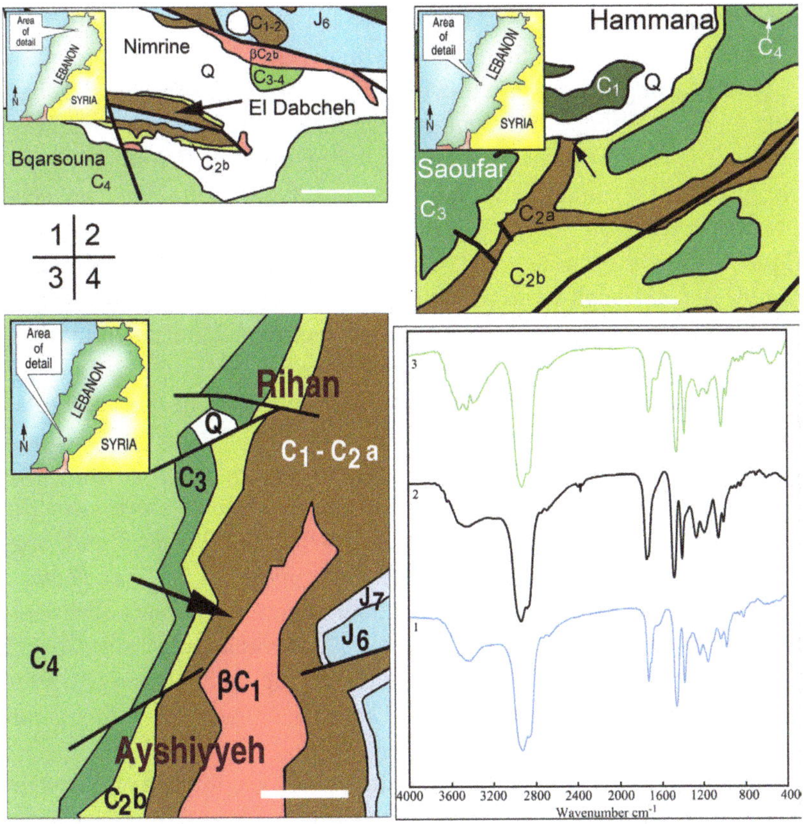

Figures 1–4. (1) Geological map of El Dabche locality: J6, Kimmeridgian; C1–2, Neocomian/Lower Aptian; C2b, Late Aptian; βC2b, Volcanic Aptian; C3–4, Albian/Cenomanian; C4, Cenomanian; Q, Quaternary; (2) Geological map of Hammana locality: C1, Neocomian; C2a, Lower Aptian; C2b, Late Aptian; βC2, Volcanic Aptian; C3, Albian; C4, Cenomanian; Q, Quaternary; (3) Geological map of Rihan locality: J6, Kimmeridgian; J7, Tithonian; βC1= Volcanic Neocomian ; C1/C2a, Neocomian/Lower Aptian; C2b, Late Aptian; C3, Albian; C4, Cenomanian; Q, Quaternary; thickened lines represent faults. Scale bar = 1 km; (4) Comparative diagram of FT-IR spectra of amber: 1, spectrum of the amber from the El Dabche outcrop; 2, spectrum of the amber from the Hammana outcrop; 3, spectrum of the amber from the Rihan outcrop.

maybe important difference can be detected between the three spectra, and it is located in the so called 'signature region' at 900 cm^{-1} where the peak indicates OH deformation in acid. The Rihan spectrum presents high peak intensity at this wavelength compared to the other two spectra. This could be due to the taphonomical process of the resin in the Rihan region that is found in sandstone sediments while the two remaining outcrops are found in dark clay sediment (rich with pyrite).

Botanical origin determination is problematic and nearly always debatable for most ambers, and sometimes different interpretations could be given for the same material depending on the workers. The botanical origin of Lebanese amber is not an exception. Trees of the Araucariaceae (*Agathis levantensis* Poinar et Milki, 2001 and *Araucarioxylon* sp.) or Cheirolepidiaceae families (*Protocarpoxylon* sp.) have both been proposed in the past, although a relationship with resin-exuding Cycadaceae or a new unknown family were also suggested (Poinar & Milki 2001; Koteja & Azar 2008; Azar et al. 2010).

The pieces of amber containing the inclusions had been prepared as proposed by Azar (2000) and Azar et al. (2003), i.e., dissected, ground, polished, and embedded in Canada balsam. The Olympus SZH10 and SZX9 stereoscopic microscopes equipped with a camera lucida were used for microscopic examination with changeable, direct and transmitted, normal and polarized, light as well as light microscope Olympus BX43. Feature measurements are considered approximate due to the effects of shrivelling and possible optical deformations. Photographs were taken using Olympus Camedia C-5060 attached to the Olympus SZH10 and Olympus BX43 microscopes, and adjusted using Combine ZP and Adobe® Photoshop Elements 6.0.

Venation terminology follows the interpretations proposed by Szwedo & Żyła (2009) and Nel et al. (2012). The following abbreviations are used: *pR*, prenodal veinlets; Pc+CP, precosta + costa posterior; CA, costa anterior; Sc+R, subcosta (posterior) + radius; Sc+RA, subcosta (posterior) + radius anterior; $ScRA_1$, terminal subcosta (posterior) + first radius anterior; RA_2, terminal second radius anterior; RP, radius posterior; M+CuA, common stalk media + cubitus anterior; CuA, cubitus anterior; CuP, cubitus posterior; Pcu, postcubitus; A_1, first anal vein; cell C1, cell between branchings of Sc+RA and RP; cell C5, cell between branchings of CuA_1 and CuA_2.

Systematics

Order Hemiptera Linnaeus, 1758
Suborder Fulgoromorpha Evans, 1946
Superfamily Fulgoroidea Latreille, 1807
Family Perforissidae Shcherbakov, 2007
Subfamily Cixitettiginae Shcherbakov, 2007

Genus *Aafrita* Szwedo et Azar gen. n.

Type species

Aafrita biladalshama sp. n., designated here.

Diagnosis

This new genus differs from *Iberofoveopsis* Peñalver et Szwedo, 2010 by the combination of characters: RP forked apicad of nodal line (RP single in *Iberofoveopsis*); stem M not forked apicad of nodal line (stem M forked basad of nodal line in *Iberofoveopsis* Peñalver et Szwedo, 2010); stem CuA forked (as in *Foveopsis* Shcherbakov, 2007 and *Iberofoveopsis*), but forking placed apicad of claval veins junction (at level of claval veins junction in *Foveopsis* and *Iberofoveopsis*); tegmen 3 times as long as wide (3.3 times as long as wide in *Iberofoveopsis*); tibio-metatarsal formula 9(9): 8(8): 7(7) (tibio-metatarsal formula 9(9): 7(7): 7(7) in *Iberofoveopsis*; 7(7): 7(7): 5(5) in *Foveopsis*); gonoplac tapered in apical portion, curved dorsad (gonoplac obliquely ending at apex in *Iberofoeopsis*).

Etymology

The generic name is derived from the word "aafrita" – the feminine form of the name of the fire demon in Arabian mythology. Gender: feminine.

Aafrita biladalshama Szwedo et Azar sp. n. (Figs 5–30)

Description

Total length about 3.5 mm. Head with compound eyes 1.09 mm wide, slightly wider than pronotum. Vertex partly covered by the anterior margin of the pronotum, visible portion 0.08 mm long at mid line, total length at mid line 0.22 mm; anterior margin of vertex carinate, lateral margins of vertex carinate, slightly converging towards frontoclypeal suture. Face subrectangular, frons 0.61 mm long at midline, 0.6 mm wide, with median carina distinct, reaching the transverse carination above the frontoclypeal suture; lateral margins carinate, upper portion with two subparallel carinae, lower carinae shallowly arcuate; with a few sensory pits at each side between median carina and lateral margin. Postclypeus 0.81 mm long, swollen, with median carina and lateral carinae in lower 3/4; anteclypeus

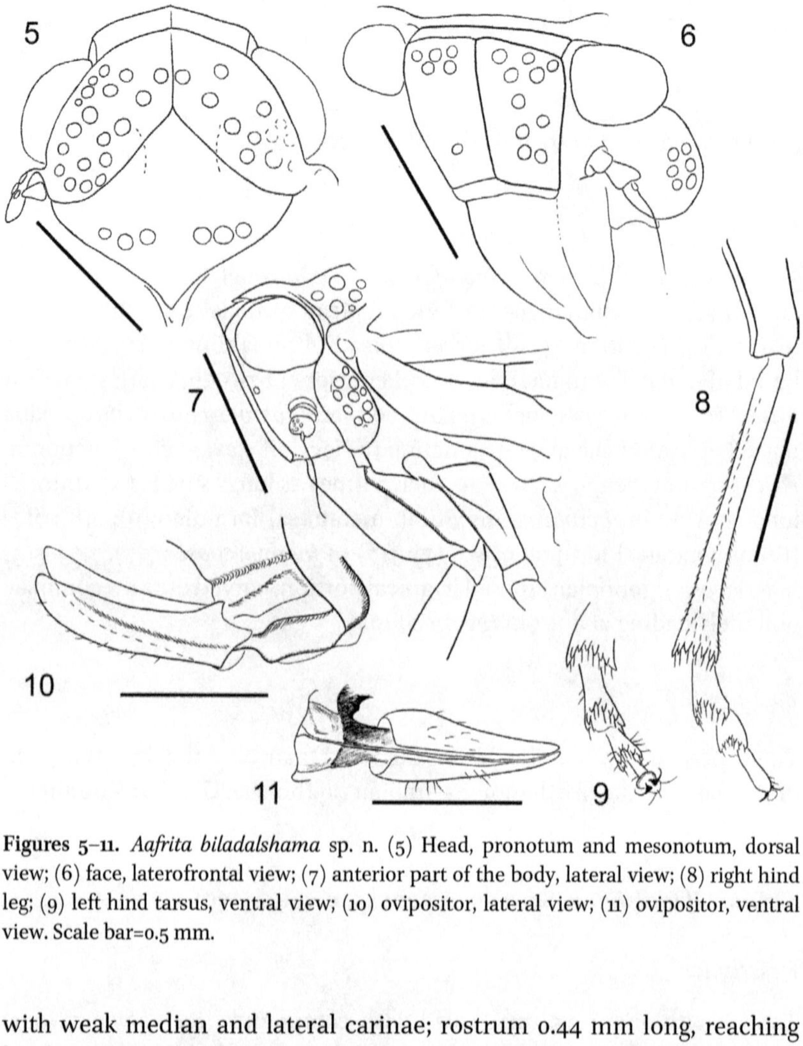

Figures 5–11. *Aafrita biladalshama* sp. n. (5) Head, pronotum and mesonotum, dorsal view; (6) face, laterofrontal view; (7) anterior part of the body, lateral view; (8) right hind leg; (9) left hind tarsus, ventral view; (10) ovipositor, lateral view; (11) ovipositor, ventral view. Scale bar=0.5 mm.

with weak median and lateral carinae; rostrum 0.44 mm long, reaching hind coxae, apical segment shorter (0.18 mm) than penultimate one, about as wide as long. Compound eyes bulging, flattened dorsoventrally. Ocelli distinct, close to the lateral carinae of face. Antennal foveae elevated, scape short, cylindrical; pedicel subconical, slightly bulbous; flagellum about 0.38 mm long.

Pronotum 0.59 mm long, 0.27 mm long at mid line, 1.02 mm wide; with anterior margin semicircular, distinctly produced between eyes over posterior head margin; overlapped by eyes laterally; with deep acutangular incision posteriorly; carinate along anterior and posterior margins; median carinae distinct, elevated in posterior half; each half of disc covered with

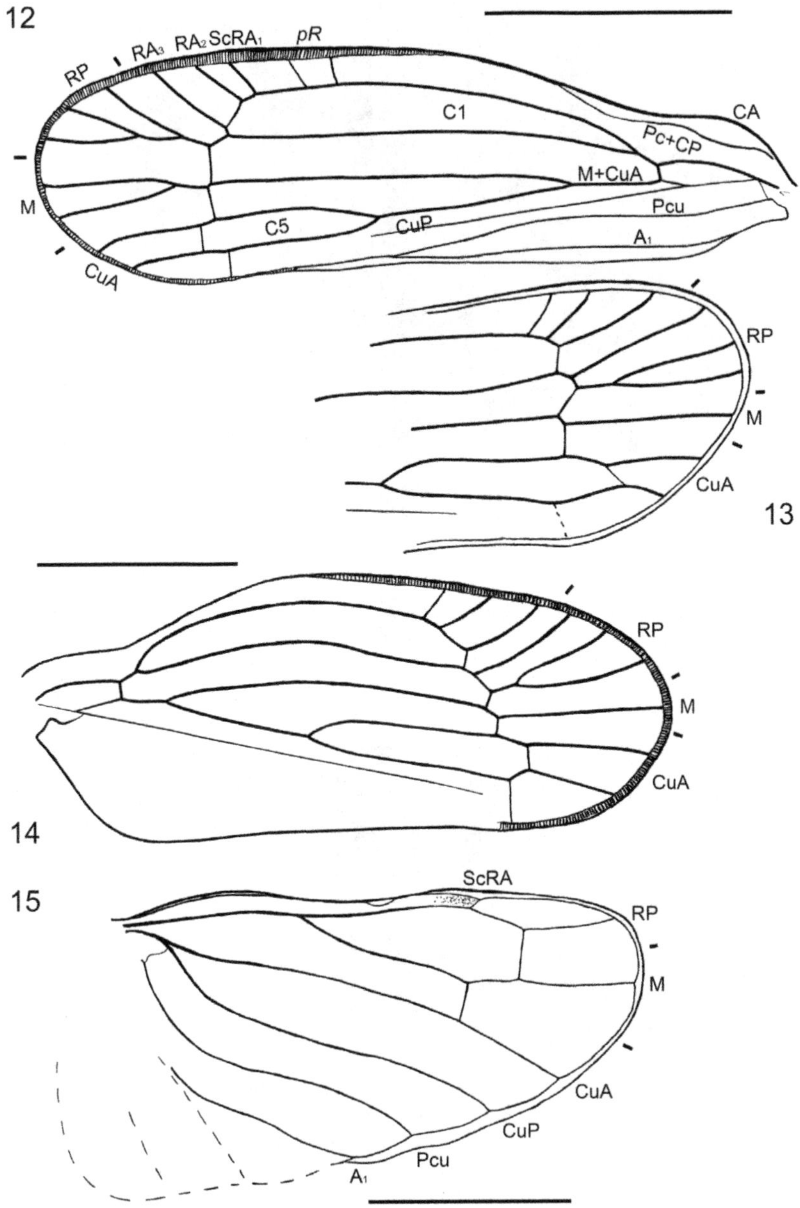

Figures 12–15. *Aafrita biladalshama* sp. n. (12) Left tegmen; (13) apex of right tegmen of the holotype; (14) left tegmen of paratype; (15) right hind wing of paratype. Scale bar=1 mm.

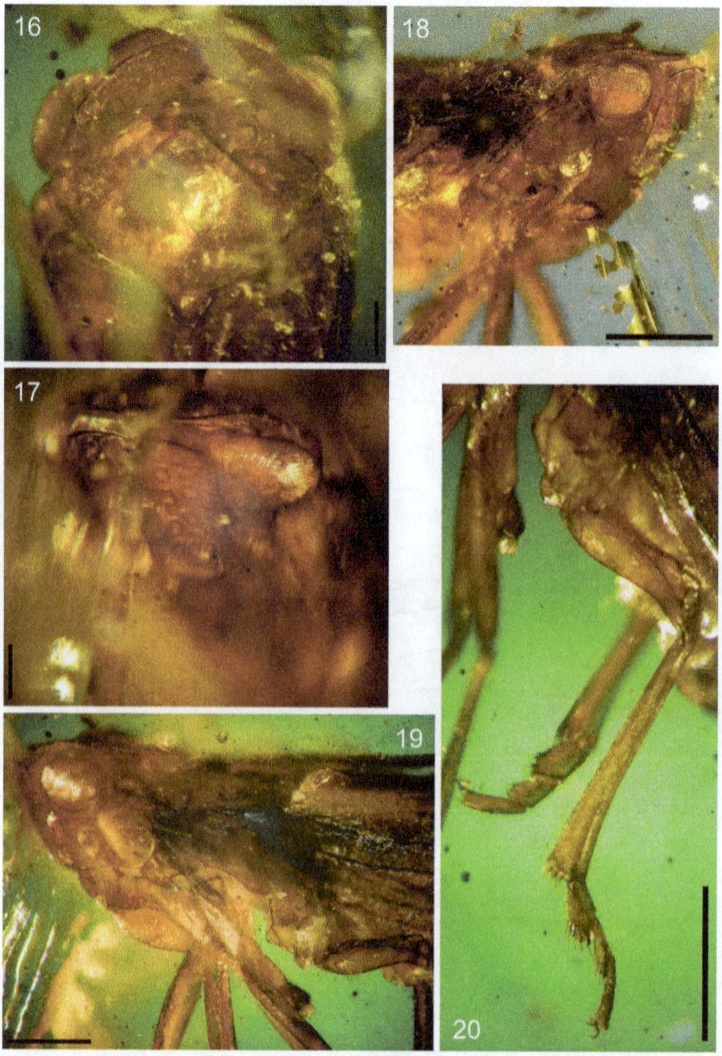

Figures 16–20. *Aafrita biladalshama* sp. n., holotype. (16) Anterior portion of body, dorsal view; (17) face, laterofrontal view; (18) anterior portion of body, right lateral view; (19) anterior portion of body, left lateral view; (20) left hind leg. Scale bar=0.25 mm for 16 and 17, 0.5 mm for 18–20.

19 sensory pits of different size (most of larger pits situated in median portion of the lobe, posteromedian portion devoid of pits); pectoral lobe posteriorly with 9 sensory pits at small convexity.

Mesonotum 0.66 mm long at mid line, 0.94 mm wide, trapezoid, convex, with three sensory pits at each side in posterolateral portions;

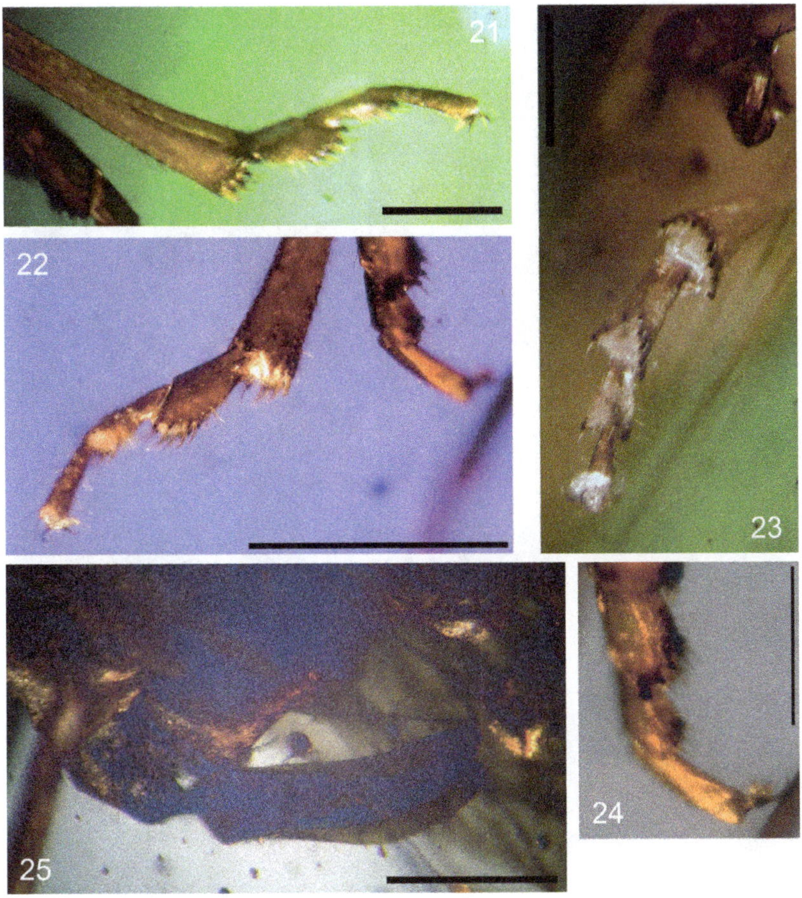

Figures 21–25. *Aafrita biladalshama* sp. n., holotype. (21) Left hind tarsus; (22) left hind tarsus; (23) right hind tarsus, ventral view; (24) left hind tarsus; (25) ovipositor. Scale bar=0.25 mm for 21, 23 and 24, 0.5 mm for 22 and 25.

mesoscutellum not separated, smooth, with apex angulately rounded. Mesepisternum with carinate eminence below the pectoral lobe of pronotum; metepisternum with carinate margin below

Tegula small, merely carinate.

Tegmen about 3-3.1 mm long, about 1.05 mm wide, coloration dark brown, lighter in apical portion; membrane finely granulate, margin of tegmen from claval apex to apex of costal area transversely wrinkled. Costal margin widely angulate at base, slightly convex above the metepisternal carination, then delicately arcuate; clavus open with claval apex at about 2/3 of tegmen length; costal margin foliaceous at base, with short

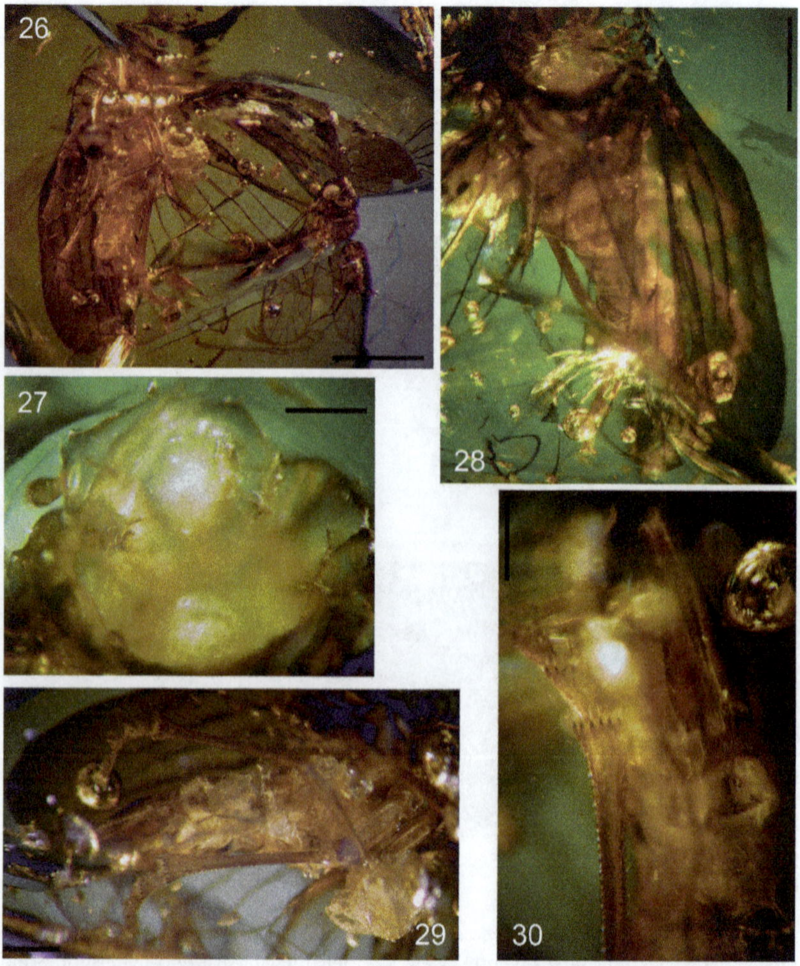

Figures 26–30. *Aafrita biladalshama* sp. n., paratype. (26) General ventral view; (27) anterior portion of body, preserved part; (28) right tegmen and wing; (29) hind legs and abdomen, ventral view; (30) right hind tarsus and ovipositor. Scale bar=1 mm for 26, 0.5 for 28; 0.25 mm for 27, 29 and 30.

costal area developed; basal cell about 4.5 times as long as wide; stem Sc+R leaving basal cell with a short stalk, branch Sc+RA carinate, arcuately curved, subparallel to costal margin, slightly more curved mediad in apical portion; two terminals, $ScRA_1$ and RA_2, reaching margin; two prenodal veinlets between ScRA and costal margin; stem RP slightly curved, forked at level of transverse veinlet ir and r-m of nodal line, reaching margin with 3(4) terminals; common stalk of M+CuA about twice as long as stem Sc+R;

stem M forked from common stalk with CuA in basal 1/3 of tegmen length, forked apicad of nodal line; stem CuA shorter than basal cell, forked slightly apicad of claval veins junction. Claval veins Pcu and A_1 fused slightly basal of half of tegmen length, common portion Pcu+A_1 shorter than free portion of Pcu. Transverse veinlet *ir* placed slightly basally of branch RP forking; veinlet *r-m* little more apicad; veinlet *m-cu* little more apicad than *r-m*, veinlet *icua* little more apicad that *m-cua*; veinlet *icu* between CuA_2 and tornus (postclaval margin) missing. Veins A_1, Pcu, stem CuA and branchings CuA_2 and CuA_1, and to a lesser degree stem M and transverse veinlets pustulate (with at most microscopic setae inserted in pustulae). Cell C5 about half as long as cell C1.

Note. Variability is observed in the postnodal portion of the tegmina. In the holotype stem M on the right tegmen is not forked, and branchings of CuA are curved and shifted mediad, with transverse veinlet *icua* distinctly oblique and shifted posterad of nodal line. In the left tegmen of paratype, stem M is not forking, branch RA_3 is forked apicad of nodal line, branch RP is forked once; veinlets *r-m* and *m-cua* are placed at same level.

Hind wing membranous, with ambient vein distinct; stem Sc+R forked apicad of wing coupling apparatus, single branch ScRA reaching ambient vein distinctly basad of wing apex, single branch RP reaching ambient vein slightly above apex; stem M separated from common stalk at the base of wing, single up to ambient vein; stems CuA, CuP and Pcu single, reaching ambient vein; anal lobe with (forked?) vein A_1, jugal lobe wide.

Fore leg with coxa elongate, about 0.42 mm long, carinate anteriorly; fore femur about 0.6 mm long, subquadrate in cross section, with ventral margins carinate; fore tibia about 0.75 mm long, slightly flattened in basal portion, subquadrate at cross section, margins carinate, covered with short setae, comb of a few setae at apex on ventral margin; fore tarsus with tarsal claws about 0.41 mm; basitarsomere 0.12 mm long, mid and apical tarsomeres subequal, 0.19 mm long; tarsal claws narrow; arolium triangularly wide, with rounded apical margin.

Mid leg with coxa about 0.4 mm long, carinate frontally; mid femur about 61 mm long, subquadrate in cross section, ventral margins slightly carinate, covered with short setae, mid tibia about 0.75 mm long, slightly widened and flattened toward apex, mid tarsus with tarsal claws about 0.41 mm; basitarsomere 0.12 mm long; mid and apical tarsomeres subequal, 0.19 mm long; tarsal claws narrow; arolium triangularly wide, with rounded apical margin.

Hind coxae close to each other; hind coxa about 0.37 mm long, with small meracanthus, carinate anterolaterad; trochanter annular, distinctly shorter than wide; hind femur about 0.53 mm long, subquadrate in cross section.; hind tibia about 0.94 mm long, subquadrate in cross section, widened towards apex, with apical row of 9 teeth bearing subapical setae; hind tarsus with tarsal claws about 0.56 mm; basitarsomere about 0.25 mm, with deep, triangular median incision, bearing 8 apical teeth with subapical setae; mid tarsomere about 0.18 mm long, with median triangular incision, bearing 7 apical teeth with subapical setae; apical tarsomere with tarsal claws about 0.31 mm long; tarsal claws narrow; arolium triangular with widely rounded lower margin.

Abdomen short. Ovipositor of orthopteroid, piercing-sawing type, about 1 mm long; with gonocoxa VIII exposed; triangular, basal portion of gonapophysis VIII exposed; gonocoxa IX partly exposed; gonoplacs elongate, in form of half-pipes surrounding the gonapophyses, with small, triangular process at base on dorsal margin; ventral margin and apical portion of gonoplacs covered with long setae; apex of gonoplacs curved dorsad, acutely rounded.

Nymphs and exuvia (Figs 31–39)

Among the inclusions three nymphs and two exuvia were found. They are provisionally associated with the new genus *Aafrita* gen. n., probably conspecific with the species *A. biladalshama* sp. n. described above. Their preservation is not very good, but some features could be observed. The nymphs probably present 1st or 2nd instars, suggested by their size, about 1.2–1.3 mm long, which is about 1/3 of the size of adult planthoppers and, development of wing pads and tarsi. Body elongate, not depressed; head with compound eyes slightly wider than pronotum. Vertex with arcuate anterior margin, with acutely rounded apex, lateral margins subparallel, posterior margin arcuately incised; vertex in mid line about as long as pronotum in mid line; anterior margin of vertex and lateral margins carinate. Frons with two carinae at upper portion; upper one at frontoclypeal ridge and lower one connected with median carina; lower carina slightly arcuate; median carina prominent; lateral margins carinate; lower margin of frons carinate, cutting off a narrow space between carina and frontoclypeal suture. Each half of frons with 10 sensory pits, 6 in upper portion and 4 in lower portion. Clypeus convex. Compound eyes bulging, oval. Scape short, pedicel subconical with apical bulbosity, base of flagellum swollen ovoid. Rostrum reaching hind coxae. Pronotum trapezoidally produced between compound eyes; median portion of anterior margin almost straight, lateral portions distinctly diverging posteriad, posterior margin

Figures 31–36. Nymphs and exuvia. (31) Specimen 623E, dorsal view; (32) specimen 623E, ventral view; (33) specimen 1156, dorsal view; (34) specimen DAB-2A, ventral view; (35) specimen DAB-2A, hind leg; (36) specimen 875, ventral view. Scale bar: 0.25 mm.

acutangularly incised anteriad; median carina (ecdysial line) distinct; each half with 10 (?) sensory pits; pectoral lobe with a group of 4 (?) sensory pits. Mesonotum with wingpads subtriangular, with shallow incision at posterior margin; mesonotum about 1.5 times as long as pronotum in mid

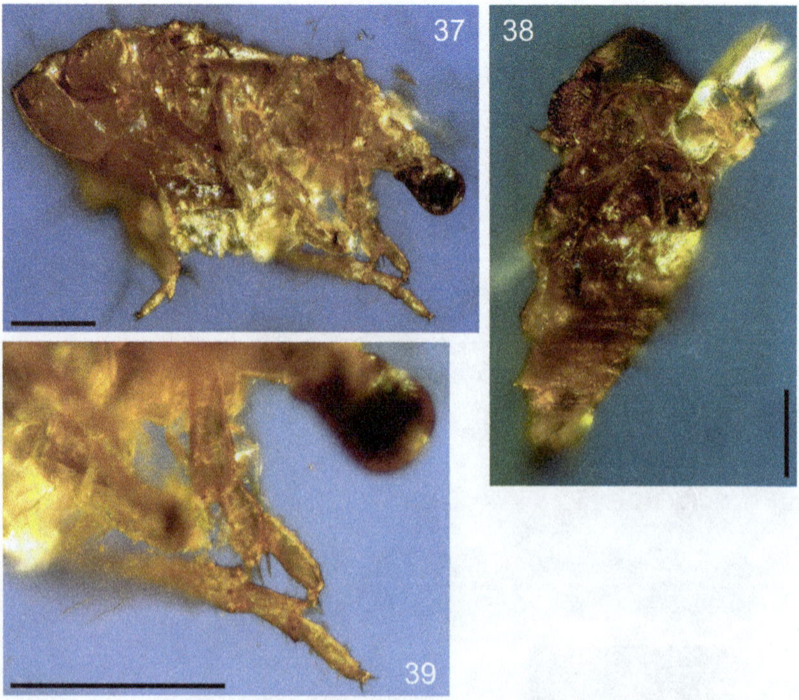

Figures 37–39. Nymph, specimen RIH-1D. (37) Ventrolateral view; (38) dorsal view; (39) hind tarsi. Scale bar=0.25 mm.

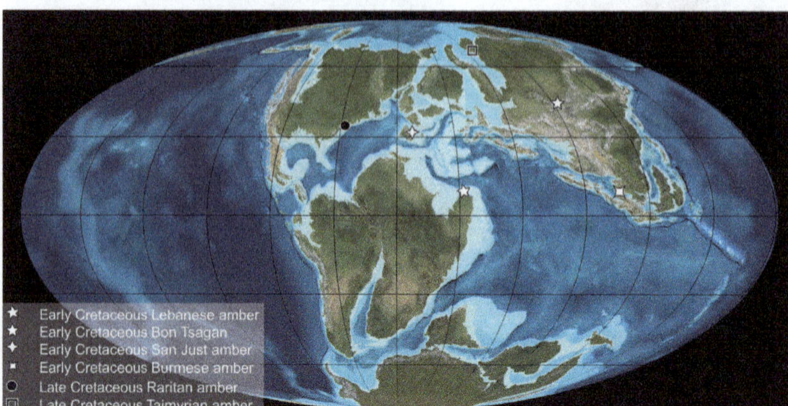

Figure 40. Distribution of the Perforissidae on the Earth as it looked 120 million years ago; palaeogeographic reconstruction courtesy of Prof. Ron Blakey (Colorado Plateau Geosystems, Boulder, CO, USA).

line; median carina (ecdysial line) distinct. Metanotum with triangular incision posteriorly, distinct median carina (ecdysial line), about half as long in mid line as mesonotum in mid line. Fore and mid legs short; tibiae with sparsely distributed short setae; tarsi two-segmented, basitarsomere slightly shorter than second tarsomere; tarsal claws distinct; arolium triangularly lobate. Hind leg with tibia longer than femur, with row of 5 apical teeth and subapical setae. Tarsi two-segmented, tarsomeres subequal in length; basitarsomere with apical row of 5 teeth and subapical setae; tarsal claws narrow; arolium triangularly lobate. Abdomen telescoped, conical in dorsal view, abdominal sensory pits not visible; 10th segment (anal tube) cylindrical, about as long as wide, distinctly longer than preceding segment.

Material examined

All the specimens examined are provisionally deposited in the Museum National d'Histoire Naturelle, Paris, and from the Dany Azar collection. Holotype, female, No. 641B, Hammana/Mdeirij locality (Central Lebanon); Paratype, female, No. 171A — female, distorted specimen, with head, pronotum and mesonotum missing, syninclusions: 2 specimens of Psocoptera (171B-C) Hammana/Mdeirij locality (Central Lebanon).

Additional material

Specimen No. 1156 — nymph, 1st or 2nd instar, Hammana/Mdeirij locality, No. 875, exuvium, partly preserved anterior portion, Hammana/Mdeirij locality; No. 623E – nymph, 1st or 2nd instar, Hammana/Mdeirij locality; No. DAB-2A — exuvium, El-Dabsheh outcrop (North of Lebanon); No. RIH-1D — nymph, 1st or 2nd instar, Rihan outcrop (Southern Lebanon).

Age and occurrence

Lower Cretaceous, Hammana/Mdeirij, Baabda District, Central Lebanon; El-Dabsheh, Sir El-Danniyeh District, North Lebanon and Rihan outcrop, Southern Lebanon.

Etymology

The specific epithet is derived from "Bilād al-Shām", a geographic and cultural term referring to the region of the eastern Mediterranean littoral between Anatolia and Egypt (Harris 2005).

Discussion

The new genus and species described above is the first report of the presence of Perforissidae in the Lower Cretaceous amber of Lebanon. *Aafrita* gen. n. presents all the family characters mentioned among the other genera of the family (Shcherbakov 2007a). It is placed in the subfamily Cixitettiginae Shcherbakov, 2007, based on the structure of the hind leg tarsal segments. Therefore the family comprises the subfamily Perforissinae, with *Perforissus muiri* Shcherbakov, 2007 from the Late Cretaceous Raritan (New Jersey) amber and *Cretargus emeljanovi* Shcherbakov, 2007 from the Late Cretaceous amber of Taïmyr Peninsula, and subfamily Cixitettiginae, with *Cixitettix yangi* Shcherbakov, 2007 from the Late Cretaceous Taïmyrian amber, *Foveopsis fennahi* Shcherbakov, 2007 from the Early Cretaceous Burmese amber, *Tsaganema oshanini* Shcherbakov, 2007 from the Early Cretaceous deposits of Mongolia and *Iberofoveopsis miguelesi* Peñalver et Szwedo, 2010 from the Early Cretaceous amber of Spain (Shcherbakov 2007a, Peñalver & Szwedo 2010). Nymphs belonging to this family were also recorded in the Early Cretaceous Jordanian amber (Kaddumi 2007). This finding gives new and important data for the palaeobiogeographic distribution of this extinct family. The finding of Perforissidae among the inclusions of the Lebanese amber is the first reliable information about presence of these insects in 'Gondwanaland' of the Early Cretaceous times (Fig. 40). It could be assumed also that representatives of this family had been widespread along both southern and northern shores of the Peritethys Sea, reaching far north to Mongolia.

As originally defined, the Perforissidae was interpreted as a neotenous offshot of the common ancestors of all recent Fulgoroidea, i.e., the Jurassic paraphyletic family 'Fulgoridiidae' (Bourgoin & Szwedo 2008). It is a peculiar group, which presents a combination of particular morphological characters not to be found among either extinct and extant Fulgoroidea. The taxonomic placement of the Perforissidae is not fully clear, as both nymphs and adults are highly derivative; however, preliminary analysis allowed the Perforissidae to be placed close to the Cixiidae and Delphacidae (Szwedo 2009; Peñalver & Szwedo 2010). The imagines of Perforissidae retain paedomorphic characters, as presence of sensory pits in adult; the destabilization and underdevelopment of hind leg armature also seems to be of neotenic nature. Reduction of number of apical teeth and subapical setae is found among various Fulgoriodea, both belonging to 'Cixiidae-like' (e.g., Derbidae) and to 'higher' Fulgoroidea (up to toothless condition in some Caliscelidae: Augilini). The tegmina are characteristic of simplified venation in comparison of supposed ancestral 'Fulgoridiidae' and probably related groups of

'Cixiidae-like' assemblage. This simplification and observed variability could also be result of paedomorphosis. The observed variability in venation on membrane in *Aafrita* gen. n. could also result from the destabilization due to neoteny. In *Iberofoveopsis* the stem M of tegmen is not forked, but slight variablitity is observed in the pattern of CuA branchings.

According to Shcherbakov's (2007a) opinion the Perforissidae were an early attempt to construct leafhopper-like forms from planthoppers, essentially quasi-leafhoppers, as a response to the colonization of the earliest angiosperms (or proangiosperms) in coastal–littoral environments.

Shcherbakov (2007a) stated that presence of multiple sensory pits in adults and nymphs indicate that the Perforissidae inhabited biotopes with highly variable and/or extreme humidity. The presence and complexity of sensory pits in the adults and nymphs of planthoppers were discussed several times (see, e.g., Emeljanov 1994, 1995; Gnezdilov & Wilson 2006, 2007; Szwedo 2007, 2009; Peñalver & Szwedo, 2010; Bräuning et al. 2012). The role of sensory pits as hygroreceptors was postulated by Šulc (1929), but it is still not supported by experimental evidence. Their other postulated function as exteroreceptors or proprioreceptors was already rejected, but their real function remains enigmatic (Bräuning et al. 2012). Then, we are still facing the apparent lack of knowledge concerning these sense organs and the understanding of the function of the sensory pits.

The Perforissidae (both adults and nymphs) were speculated to live on low herbaceous proangiosperms and/or angiosperms, and/or cheirolepidiacean gymnosperms (Shcherbakov 2007a). The relationship of Perforissidae with cheirolepidiacean gymnosperms is supported also by the findings of the early instars in amber of presumed cheirolepidiacean origin, i.e., Lebanese amber. Ponomarenko (1998) considered that the transformation of the proangiosperms into angiosperms (angiospermization) has been ruled by neoteny. The Perforissidae, the presumed trophic associates of these plants, very probably also have been subjects of this rule. The reasons for extinction of this family of planthoppers could be the strong competition from the Cicadellidae (Shcherbakov & Popov 2002) or close trophic relationships with groups of early (pro)angiosperms or gymnosperms which did not survive into the Cenozoic (Szwedo 2009).

Acknowledgements

This paper is a contribution to the project "Insects in the Lower Cretaceous Lebanese amber: taxonomy, palaeobiodiversity and evolution", under cooperative agreements between the Lebanese University (Beirut); MIZ

(Museum and Institute of Zoology), Polish Academy of Sciences, Warsaw; University of Gdańsk; and the Polish Academy of Sciences, Krakow project "Origin and evolution of biodiversity of Europe and the Middle East". This paper is a contribution to the team project "Biodiversity: Origin, Structure, Evolution and Geology" awarded to D.A. by the Lebanese University. Y.N. would like to thank the AZM and SAADE Association for partially financing his PhD, and the PRASE in the Lebanese University for allowing geochemical analyses.

References

Azar, D. (2000) *Les ambres mésozoïques du Liban*. Thèse de Doctorat de l'Université Paris XI, Paris, 164 pp. +144 pp. annexes.

Azar, D. (2012) Lebanese amber: a "Guinness Book of Records". *Annales Universitatis Paedagogicae Cracoviensis, Folia* **111**: 44–60.

Azar, D., Perrichot, V., Néraudeau, D. & Nel, A. (2003) New psychodid flies from the Cretaceous ambers of Lebanon and France, with a discussion about *Eophlebotomus connectens* Cockerell, 1920 (Diptera, Psychodidae). *Annals of the Entomological Society of America* **96**: 117–127.

Azar, D., Gèze, R. & Acra, F. (2010) Chapter 14: Lebanese amber. In: Penney, D. (Ed.) *Biodiversity of Fossils in Amber*. Siri Scientific Press, Manchester, pp. 271–298.

Bourgoin, Th. (2013) *FLOW (Fulgoromorpha Lists on The Web): a world knowledge base dedicated to Fulgoromorpha*. Version 8, updated 23 January 2013. Available online at http://hemiptera-databases.org/flow/ (accessed 29 January 2013).

Bourgoin, Th. & Szwedo, J. (2008) The 'cixiid-like' fossil planthopper families. *Bulletin of Insectology* **61**: 107–108.

Bräunig, P, Krumholz, K. & Baumgartner, W. (2012) Sensory pits – Enigmatic sense organs of the nymphs of the planthopper *Issus coleoptratus* (Auchenorrhyncha, Fulgoromorpha). *Arthropod Structure & Development* **41**: 443–458.

Emeljanov, A.F. 1994 (1993). Morphological peculiarities of larvae of dictyopharid planthoppers (Homoptera, Dictyopharidae). I. General characteristics and identification keys to genera of the Palaearctic fauna. *Entomologicheskoe Obozrenie* 72: 794–819. (In Russian)

Emeljanov, A.F. (1995) On the problem of classification and phylogeny of the family Delphacidae (Homoptera, Cicadina) taking into consideration larval characters. *Entomologicheskoe Obozrenie* **74**: 780–794. (In Russian)

Gnezdilov, V.M. & Wilson, M.R. (2006) Systematic notes on tribes in the family Caliscelidae (Hemiptera: Fulgoroidea) with the description of new taxa from Palaearctic and Oriental Regions. *Zootaxa* **1359**: 1–30.

Gnezdilov, V.M. & Wilson, M.R. (2007) A new genus and a new species of the tribe Mithymnini (Hemiptera: Fulgoromorpha: Nogodinidae) from Namibia, with sternal sensory pits in the adult. *Zootaxa* **1453**: 55–62.

Hamilton, K.G.A. (1990) Homoptera. *Bulletin of the American Museum of Natural History* **195**: 82–122.

Harris, W.W. (2005) *The Levant: a fractured mosaic*, 2nd edition. Markus Wiener, Princeton, NJ, 218 pp.

Kaddumi, H.F. (2007) *Amber of Jordan: the oldest prehistoric insects in fossilized resin*, 3rd edition. Eternal River Museum of Natural History, Amman, 268 pp.

Koteja, J. & Azar, D. (2008) Scale insects from Lower Cretaceous amber of Lebanon (Hemiptera: Sternorrhyncha: Coccinea). *Alavesia* **2**: 133–167.

Lin, Q.-b., Szwedo, J., Huang, D.-y. & Stroiński, A. (2010) Weiwoboidae fam. nov. of 'higher' Fulgoroidea (Hemiptera: Fulgoromorpha) from the Eocene deposits of Yunnan, China. *Acta Geologica Sinica* (English Edition) **84**: 751–755.

Nel, A., Prokop, J., Nel, P., Grandcolas, P., Huang, D.-y., Roques, P., Guilbert, E., Dostál, O. & Szwedo, J. (2012) Traits and evolution of wing venation pattern in paraneopteran insects. *Journal of Morphology* **273**: 480–506.

O'Brien, L.B. (2002) The wild wonderful world of Fulgoromorpha. In: Holzinger, W. (Ed.) Zikaden – leafhoppers, planthoppers and cicadas (Insecta: Hemiptera: Auchenorrhyncha), zugleich Kataloge des OÖ. Landesmuseums, Neue Folge Nr. 176. *Denisia*, **4**: 83–102.

Peñalver, E. & Szwedo, J. (2010) Perforissidae (Hemiptera: Fulgoroidea) from the Cretaceous San Just amber (Eastern Spain). *Alavesia* **3**: 97–103.

Poinar Jr., G.O. & Milki, R. (2001) Lebanese amber: the oldest insect ecosystem in fossilized resin. Oregon State University Press. Corvallis, OR, 96 pp.

Ponomarenko, A.G. (1998) Paleobiology of angiospermization. *Paleontological Journal* **32**: 325–331.

Shcherbakov, D.E. & Popov, Yu.A. (2002) Superorder Cimicidea Laicharting, 1781 Order Hemiptera Linné, 1758. The bugs, cicadas, plantlice, scale insects, etc. (= Cimicida Laicharting, 1781, = Homoptera Leach, 1815 + Heteroptera Latreille, 1810). In: Rasnitsyn, A.P. & Quicke, D.L.J. (Eds) *History of Insects*. Kluwer, Dordrecht, pp. 143–157.

Shcherbakov, D.E. (2007a) An extraordinary new family of Cretaceous planthoppers (Homoptera: Fulgoroidea). *Russian Entomological Journal* **16**: 139–154.

Shcherbakov, D.E. (2007b) Mesozoic spider mimics – Cretaceous Mimarachnidae fam. n. (Homoptera: Fulgoroidea). *Russian Entomological Journal* **16**: 259–264.

Szwedo, J. (2007) Nymphs of a new family Neazoniidae fam. n. (Hemiptera: Fulgoromorpha: Fulgoroidea) from the Lower Cretaceous Lebanese amber. *African Invertebrates* **48**: 127–143.

Szwedo, J. (2008) Distributional and palaeoecological pattern of Lower Cretaceous Mimarachnidae (Hemiptera: Fulgoromorpha). *Entomologia Generalis* **31**: 231–242.

Szwedo, J. (2009) First discovery of Neazoniidae (Insecta, Hemiptera, Fulgoromorpha) in the Early Cretaceous amber of Archingeay, SW France. *Geodiversitas* **31**: 105–116.

Szwedo, J. & Żyła, D. (2009) New Fulgoridiidae genus from the Upper Jurassic Karatau deposits, Kazakhstan (Hemiptera: Fulgoromorpha: Fulgoroidea). *Zootaxa* **2281**: 40–52.

Szwedo, J., Bourgoin, Th. & Lefebvre, F. (2004) *Fossil Planthoppers (Hemiptera: Fulgoromorpha) of the World. An annotated catalogue with notes on Hemiptera classification*. Studio 1, Warsaw, pp. 1–199+8 Pls.

Szwedo, J., Wang, B. & Zhang, H.-c. (2011) An extraordinary Early Jurassic planthopper from Hunan (China) representing a new family Qiyangiricaniidae fam. nov. (Hemiptera: Fulgoromorpha: Fulgoroidea). *Acta Geologica Sinica* (English Edition) **85**: 739–748.

Šulc, K. (1928) Die Wachsdrüse und ihre Produkte bei den Larven der Cixiinen (Homoptera). *Biologické Spisy Vysoké Školy Zvěrolékařské*, Brno 7 (3) [= sign. B 108]: 1–32.

PART FOUR

FOSSIL INSECT COMPRESSION

Mieroblattina pacis gen. et sp. n. – Upper Cretaceous cockroach (Blattida: Mesoblattinidae) from Nammoura limestone of Lebanon

Peter Vršanský[a,b] and Edouard Makhoul[c]

[a]Geological Institute, Slovak Academy of Sciences, Dúbravská cesta 9, 840 05 Bratislava 45, Slovakia; e-mail: geolvrsa@savba.sk
[b]Arthropoda Laboratory, Paleontological Institute, Russian Academy of Sciences, Profsoyuznaya 123, 117868 Moscow, Russia
[c]Saint-Joseph University, P.O. Box 165507, Beirut, Lebanon
e-mail: maked@cyberia.net.lb

Abstract

Mieroblattina pacis gen. et sp. n. based on an adult, possibly brachypterous or apterous female is described from the Cenomanian sediments of Nammoura in Lebanon, representing the last known Eurasian and/or African occurrence of the family Mesoblattinidae. It appears derived from taxa close to *Hispanoblatta* Martínez-Delclòs, 1993.

Keywords

Insecta; Blattaria = Blattodea; Mesoblattinidae; new genus; Cenomanian; Upper Cretaceous; Lebanon

Introduction

The Mid Cenomanian "fish beds" of Nammoura in Lebanon contain a rich and diverse fauna of marine fishes, aquatic and terrestrial reptiles, bird remains, small crustaceans (Schram et al. 1999), cephalopods, coprolithes and small faecal pellets. A well-preserved plant assemblage contains aquatic ferns, gymnosperms, and angiosperm flowers and leaves (Krassilov & Bacchia 2000). Insects are very rare in this "lithographic" limestone. Based on four complete specimens, Nel et al. (2004) described *Chresmoda libanica*, the latest known representative of the enigmatic insect family Chresmodidae, and an undetermined beetle is also recorded. Recently, Tony Dakkash found the cockroach specimen described herein.

Systematic Palaeontology

Order Blattida Latreille, 1810
Family Mesoblattinidae Handlirsch, 1906

Mieroblattina Vršanský & Makhoul gen. n.

Type species

Mieroblattina pacis sp.n., by monotypy.

Diagnosis

Total body size about 23 mm. Abdomen with sophisticated coloration pattern, i.e., with coloured margins only and with coloration expanding posteriorly. Head comparatively small, covering less than half of the pronotal width.

Pronotum transversal with sophistic shape, viz. large, transversal, widest in the basal third, sharpened basally in the centre (as in Fig. 1d). Sternal segments 3–10 with parallel margins and characteristic coloration of margins, expanding posteriorly. Ovipositor short, partially internalised as in *Hispanoblatta*. Cerci long, with about 15 segments. Legs cursorial, carinated.

Etymology

Mieroblattina is a combination of mier (Slavic for peace) and the genus name *Blatta*.

Mieroblattina pacis Vršanský & Makhoul sp. n. (Fig. 1c, d)

Type material

Holotype specimen NI 5a-b (NI = Nammoura Insect, part and counterpart), collection of Edouard Makhoul, Beirut, Lebanon, provisionally housed in Muséum National d'Histoire Naturelle, Paris. Head, pronotum, body with ovipositor, and middle and hind legs of an adult female.

Type locality

Nammoura, El Ghabour valley, Caza Kesrouan, Mouhafazit Jabal Loubnan (Mont-Liban district), Lebanon (Fig. 1a).

Figure 1. (a, b) The type locality and horizon of the "peace cockroach" *Mieroblattina pacis* Vršanský et Makhoul, gen. et sp. n. (a) The outcrop of Nammoura, El Ghabour valley, Caza Kesrouane, Mouhafazit Jabal Loubnan (Mont-Liban district), Lebanon; (b) details of the outcrop. (c, d) *Mieroblattina pacis* Vršanský et Makhoul, gen. et sp. n. Holotype. NI 5a,b. (NI = Nammoura Insect, a-b: part and counterpart), collection Edouard Makhoul, Beirut, Lebanon, provisionally housed in Muséum National d'Histoire Naturelle, Paris, France. Cenomanian Upper Cretaceous, Lebanon. (c) negative (NI 5b) photographed under alcohol; (d) reconstruction (reverse).

Stratigraphic horizon

Mid-Cenomanian, Upper Cretaceous.

Description

Head width 3.5 mm. Pronotum width/length 5.5/8.5 mm; abdomen 8.6 mm wide, with distinct colouration (as in Fig. 1d). Cerci long at least 3.4 mm, with at least 14 segments, segments 3–6 very short. Each ventral side of segments with at least seven large chaetica in each row. First six segments with one row of sensilla, segments 7–9 with two rows, terminal segments with three rows of sensilla.

Mid leg femora 6.2 mm long, hind femora short, tibiae 9.6 mm long, with seven apical spines.

Etymology

Pacis is after pax (Latin for peace). Proposed vernacular name for the species is the "peace cockroach".

Remarks

The wings are missing, but the completely formed short external ovipositor indicates that it is an adult female and not a larva. Eventual brachypterous or apterous adult female would be the earliest and the only Mesozoic occurrence of any described brachypterous or apterous cockroach.

Certainly there is a possibility that wings were simply lost during the deposition, but in cockroaches, body is detached first, and such preserved specimen is unknown in dozens of thousands specimens. Therefore at least a possibility of apterous or brachypterous individual must be considered, in spite of the fact that no corresponding structures are visible on the specimen.

The rather particular shape of the pronotum, abdomen with nearly parallel both sternal and both tergal margins (taphonomical factors tend to act against the linearity of these characters), cerci with roughly 15 segments and the short ovipositor are characteristic for the family Mesoblattinidae (Vršanský & Ansorge 2007). This family gave birth to the family Ectobiidae (and is paraphyletic in this respect) and this transition is characteristically gradual—incomplete hiati are typical of cockroach evolution in all levels down to species—a consequence of the vast fossil record (Vršanský et al. 2011). It is namely this rich record, which allow us to see that twelve of 29 recognised and well defined cockroach families are paraphyletic (8 of them gave birth to more than one family), so there is no reason for synonymisation of these two taxa.

The large size, general appearance, the strongly transversal, ovoid shape of the pronotum and ovipositor (extremely short with parallel margins) are synapomorphic with *Hispanoblatta* Martínez-Delclòs, 1993, but sterna with black and white sophisticated colouration of margins are exceptional.

Archimesoblatta Vršanský, 2003 is of roughly similar size, but has straight pronotal margins (apomorphy, ancestral stem Mesozoic cockroaches have pronotum sharpened basally), and elongated wings and body (as in this family the length of body and wings correlates), unlike in the present genus.

All other genera of the Mesoblattinidae, sensu Vršanský & Ansorge (2007) differ in being considerably smaller and having slender habits.

The present large coloured species enhances the morphological variability within this, otherwise extremely morphologically uniform family.

Acknowledgements

The authors thank Mr Tony Dakkash, who found this material and made it available for study. We also thank Drs André Nel and Dany Azar for the valuable remarks and outcrop illustrations. Supported by UNESCO-AMBA and IGCP 458, VEGA 6002 and 2/0125/09, MVTS and Literárny Fond.

References

Martínez-Delclòs, X. (1993) Blátidos (Insecta, Blattodea) del Cretácico Inferior de España. Familias Mesoblattinidae, Blattulidae y Poliphagidae. *Boletin Geologico y Minero* **104**: 52–74.

Handlirsch, A. (1906) *Die fossilen Insekten und die Phylogenie der rezenten Formen, Band I.* Engelmann, Leipzig, 1340 pp.

Krassilov, V. & Bacchia, F. (2000) Cenomanian florule of Nammoura, Lebanon. *Cretaceous Research* **21**: 785–799.

Latreille, P.A. (1810) *Considérations générales sur l'ordre naturel des animaux composant les classes des Crustacés, des Arachnides et des Insectes avec un tableau méthodique de leurs genres disposés en familles.* Schoell, Paris, 444 pp.

Nel, A., Azar, D., Martínez-Delclòs, X. & Makhoul, E. (2004) A new Upper Cretaceous species of *Chresmoda* from Lebanon – a latest representative of Chresmodidae (Insecta: Polyneoptera inc. sed.): first record of homeotic mutations in the fossil record of insects. *European Journal of Entomology* **101**: 145–151.

Schram, F.R., Hof, C.H.J. & Steeman, F.A. (1999) Thylacocephala (Arthropoda: Crustacea?) from the Cretaceous of Lebanon and implications for thylacocephalan systematics. *Palaeontology* 42: 769–797.

Vršanský, P. (2003) Unique assemblage of Dictyoptera (Insecta– Blattaria, Mantodea, Isoptera) from the Lower Cretaceous of Bon Tsagaan Nuur in Mongolia. *Entomological Problems* 33: 119–151.

Vršanský, P. & Ansorge, J. (2007) Lower Jurassic cockroaches (Insecta: Blattaria) from Germany and England. *African Invertebrates* 48: 103–126.

Vršanský, P., Cifuentes-Ruiz, P., Vidlička, L., Čiampor, F.J.R. & Vega, F.J. (2011) Afro-Asian cockroach from Chiapas amber and the lost Tertiary American entomofauna. *Geologica Carpathica* 62: 463–475.

PART FIVE

NEW AMBER OUTCROP

Fanar, a "dream" Lebanese Lower Cretaceous amber outcrop, dissipated…

Dany Azar[a,*], Youssef Nohra[a,b,c], Denise Iskandar[a] and Raymond Gèze[a]

[a]Lebanese University, Faculty of Sciences II,
Department of Natural Sciences, Fanar P.O. Box 26110217, Fanar–Matn, Lebanon
[b]Lebanese University, Doctoral School of Sciences and Technologies, Campus Rafic,
P.O. Box 5, Hariri–Hadath, Lebanon
[c]Géosciences Rennes (UMR CNRS 6118), Université de Rennes 1,
Campus de Beaulieu Building 15, avenue du General Leclerc 263,
35042 Rennes cedex, France
*Corresponding author, e-mail: azar@mnhn.fr

Abstract

A new "ephemeral" Neocomian amber deposit is discovered in Fanar (Central Lebanon), a close suburb of the capital Beirut. This outcrop is described, its amber is characterized chemically, and palynological analysis of its sediments is done, allowing as such date information, and palaeoenvironment reconstruction.

Keywords

Lebanon; amber outcrop; Neocomian; Palynology; FT-IR analysis; palaeoenvironments

Introduction

Lebanon is well known for its numerous Mesozoic amber deposits. Most of the amber outcrops in Lebanon are Lower Cretaceous in age (circa 450 outcrops), but recently 19 new Jurassic outcrops have been discovered. Curiously, to date only 22 Lower Cretaceous amber localities gave biological inclusions.

Here we report the discovery of a new outcrop of amber, in Fanar, just in front of our University. The amber of this outcrop is characterized then analysed using Fourier Transform Infrared (FT-IR) spectroscopy. Palynological analysis of amber-bearing sediments is conducted, enabling date information and palaeoenvironment reconstruction.

Site History and Description

Whilst any discovery of any amber outcrop from any age and in any country takes a great mediatic dimension, a considerable number of outcrops (several dozen) have been recovered in Lebanon since few years without even a small note on their existence.

Since the beginning of the year 2011, an excavation was taking place in the Lower Cretaceous sandstone (Neocomian) layers just aside the Faculty of Sciences II of the Lebanese University in Fanar (place of our work) (Fig. 1), for the purpose of constructing a building (Fanar being a very near suburb of the city of Beirut, less than 5 km far away from the city). As soon as the excavation was progressing some dark grey layers appeared (Fig. 2). These layers were very promising for holding amber but not accessible as bulldozers and workers were very active during all the week (Fig. 3). On Saturday 22 January just after the weekend holiday begun, one of us (D.A.) went to explore the site, and he found some amber pieces and a lot of vegetal remains in dark grey lenses of clay evoking fluvial channels as an initial place of deposition (Figs 4 and 5). Maximum possible amount of material (amber, lignite, plant remains and sediments for palynology) was collected that day (in hard meteorological condition), in order to rescue at least a memory of this "short available" amber deposit.

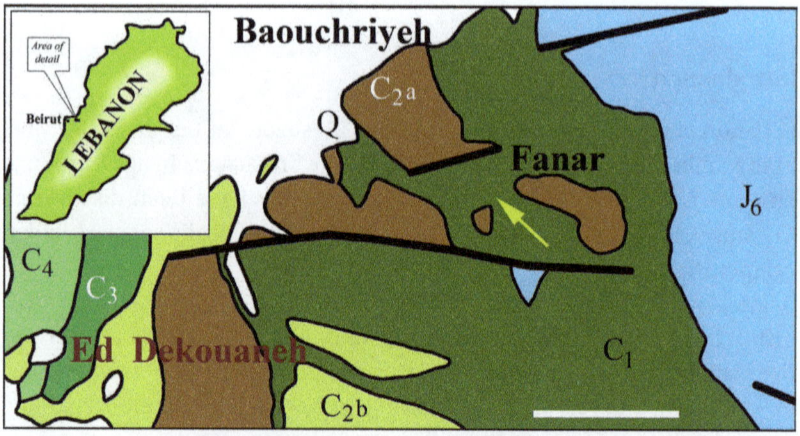

Figure 1. Geological map of amber outcrop in Fanar (adapted from Dubertret (1945). Abbreviations: J_6, Kimmeridgian; C_1, Neocomian; C_{2a}, Lower Aptian; C_{2b}, Late Aptian; C_3, Albian; C_4, Cenomanian; Q, Quaternary. Thickened lines represent faults. Scale bar=1 km.

Figure 2. Fanar outcrop.

Figure 3. The activity in the outcrop. Unfortunately, the bulldozers here are not digging for amber!

Figure 4. Amber of Fanar. Scale bar=1 cm.

Figure 5. The only real "useful" activity on the site: collecting amber.

Amber Description

The amber from this site (33°52′42.43″ N, 35°34′07.31″ E) is opaque to translucent with a range of colours varying from clear-brown to very dark orange. Amber pieces are nearly as large as beans' seeds (Fig. 4). The maximum amount of material was collected as the outcrop is no more accessible. It is noteworthy that this is the second outcrop of amber to be discovered in Fanar; the first one being 600 m to the North-West of the new one, and discovered more than 15 years ago by one of us (D.A.); it is also today lost as it is laying at present beneath a building. A regrettable fact is that the new amber deposit has also disappeared beneath a building.

FT-IR Analysis

A FT-IR spectrum was obtained of the new amber material in order to keep a chemical signature of this material, and to compare its chemical structure profile to other ambers profiles collected from Lebanon.

For the infrared analysis, 0.2 mg of amber was crushed and mixed with KBr (FT-IR grade; Merck, Darmstadt, Germany) and pellets were prepared using a manual press. Transmission FT-IR spectroscopy was performed with a Bruker IFS 55 spectrophotometer. The spectrum was acquired between 4000 and 400 cm^{-1} with 40 scans collected at 4 cm^{-1} resolution.

The spectrum demonstrates clearly that this amber is characterized by similar features as the remaining Lower Cretaceous ambers collected in Lebanon (Azar et al. 2012). The FT-IR spectrum of the amber from Fanar outcrops is shown in Fig. 6. Transmittance peaks at 3400 cm^{-1}, 2955 cm^{-1},

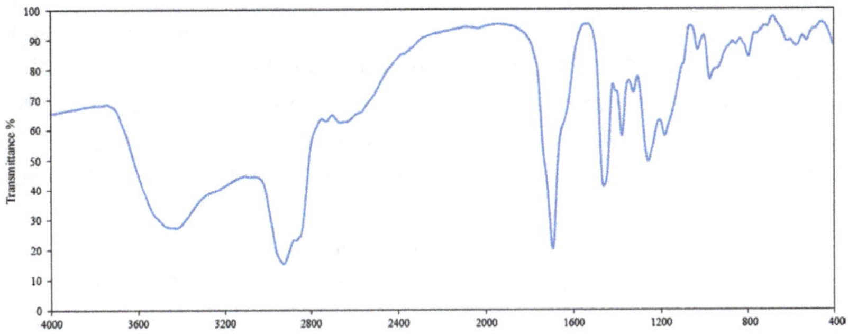

Figure 6. FT-IR spectrum of the amber of Fanar.

2865 cm^{-1} and 2925 cm^{-1} wavelength correspond to O–H stretching in phenolic and aliphatic compounds, to C–H asymmetric and symmetric stretching in CH_2 and C–H asymmetric stretching in CH_3, respectively. Carbonyl (C=O) functional groups can be detected in the range of 1725–1680 cm^{-1}. The peaks are related to the stretching vibration of the C=O in free COOH groups, esters. Also, the C=O functional groups are detected at the range of 1600–1540 cm^{-1} resulting an asymmetrical stretching in COOH groups and in carboxylate complex groups. The 1455 cm^{-1} and 1377 cm^{-1} peak points reflect the OH deformation in phenolic groups, and C–N stretching in amines. C–O–C vibrations in esters from acetate and formate, are observed at 1258 and 1182 cm^{-1}. Finally, C–OH stretching vibration in secondary cyclic alcohols appears at 1029 cm^{-1}.

Palynology of the Site

Sediments containing the amber were collected in order to carry out a palynological study of the site and deduce the paleoenvironment. These samples were prepared following the standard traditional method of extracting palynomorphs from the sediment matrix based on acid attack in a fume hood (hydrochloric acid (HCl), hydrofluoric acid (HF) over several days, nitric acid (HNO_3)). Afterwards, the attacked sediment was filtered at 8 µm. Obtained residue was then mounted on microscopic slides with a glycerin jelly (plus one drop of phenol) medium. Observation and illustrations were made with a differential interference contrast microscope.

Palynofacies

The palynofacies shows the mixture of marine and terrestrial contents. Analysed slides show dominance of marine palynomorphs (82%), especially very well preserved and abundant dinoflagellate cysts, accompanied by chitinous foraminiferal linings (Fig. 7). Amorphous carbonized (coaly) debris are rare. Nevertheless, slides are rich in terrestrial vegetal fragments (scraps of epidermis, cuticles). Sporomorphs (pollen (12%) and spores (6%)) are moderately preserved.

The reduced percentage of carbonized debris coinciding with a fairly assertive presence of pollen, spores and cuticles would reflect the development of the vegetation cover (in the amber producing forest) that protected the soil against erosion and affected its excessive drainage by water streams.

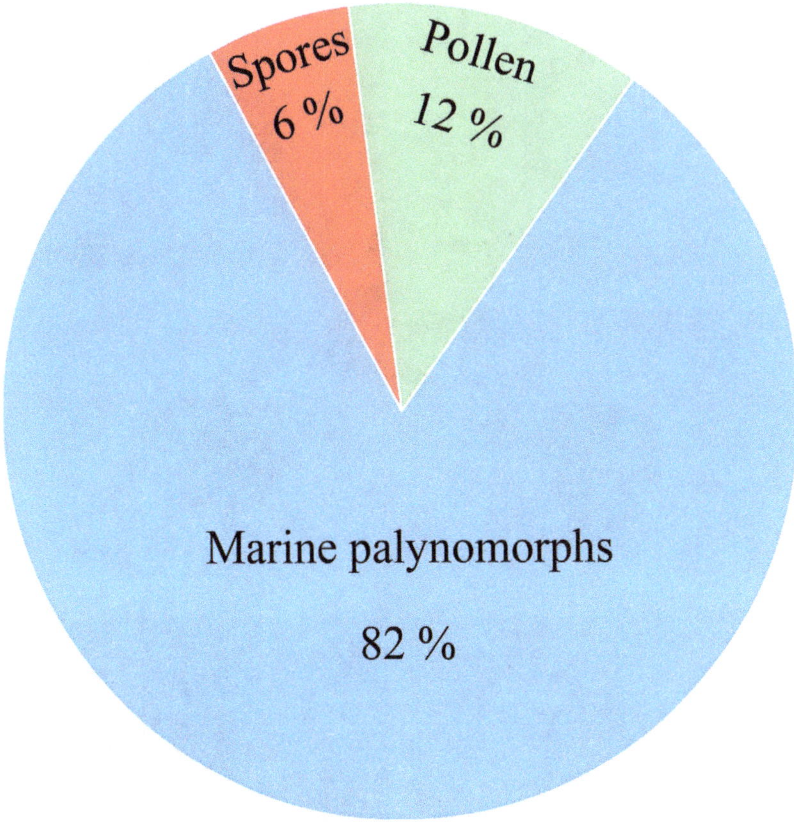

Figure 7. Diagram representing the percentage of repartition of marine and continental palynomorphs.

Marine Palynomorphs

About 17 taxa of marine palynomorphs were identified in the treated sediment (Fig. 8) (a complete list of taxa will be given in a further publication), corresponding to 82% (533) of total palynomorphs counted (647).

Among Dinoflagellata (Bütschli 1885) the genus *Oligosphaeridium* Davey & Williams, 1966 alone represents 75% of marine palynomorphs. Other genera of dinoflagellate cysts are also present, like *Spiniferites* Mantel, 1850, *Aptea* Eisenack, 1958, *Cribroperidinium* Neale & Sarjeant, 1962, and? *Cyclonephelium* Deflandre et Cookson, 1955, in addition to other genera that are present in smaller amounts. *Oligosphaeridium* and *Spiniferites* are indicators of open neritic conditions (Downie et al. 1971; Masure 1984).

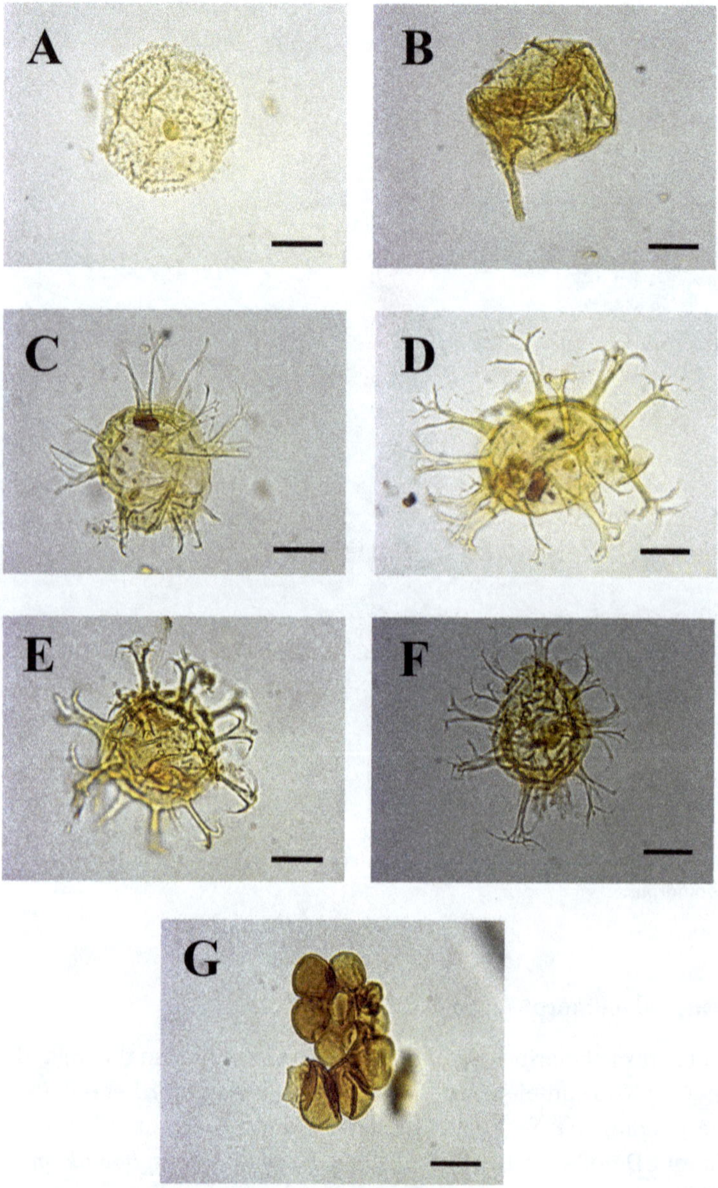

Figure 8. Marine palynomorphs. (A) *Aptea anaphrissa* (Sarjeant, 1966) Sarjeant & Stover 1978 (slide 8, S45 3); (B) *Cribroperidinium longicorne* (Downie, 1957) Lentin & Williams 1985 (slide 7, J50 0); (C) *Oligosphaeridium dividuum* Williams, 1978 (slide 4, R29 0); (D and E) *Oligosphaeridium pulcherrimum* (Deflandre et Cookson, 1955) Davey et Williams, 1966 (respectively slide 7, N52 1 and slide 4, P35 3); (F) *Spiniferites ramosus* (Ehrenberg 1838) Loeblich et Loblich, 1966 (slide 8, W37 0); (G) *Trochiliascia cuvillieri* Deák, 1964 (slide 7, O33 3). Material is deposited in the Natural History Museum of the Lebanese University.

Cyclonephelium is mainly recorded in coastal environments (Eshet et al. 1992). Chitinous foraminiferal linings are present as well, the foraminifers are generally abundant in coastal environments and shallow platform (Batten 1996).

Continental Palynomorphs

Terrestrial microflora is little diversified. Among spores the genera *Deltoidospora* (Miner, 1935) Potonié, 1956 and *Dictyophyllidites* (Couper, 1958) Dettmann 1963 are the best represented (15 and 6% of continental palynomorph assemblage, respectively) among other Pteridophyta (*Biretisporites* (Delcourt et Sprumont, 1955) Delcourt, Dettmann et Hughes, 1963; *Concavissimisporites* (Delcourt et Sprumont, 1955) Delcourt, Dettmann et Hughes, 1963; *Trilobosporites* (Pant, 1954) Potonié, 1956, *Coptospora*, Dettmann, 1963; *Cicatricosisporites* Potonié et Gelletich, 1933, among others) (Fig. 9).

In pollen *Araucariacites australis* Cookson, 1947 alone represents 28% of continental palynomorphs and is dominant in general and particularly among gymnosperm pollen. Among continental palynomorphs, other gymnosperm pollen grains are present (*Classopolis* (Pflug, 1953) Pocock et Jansonius, 1961 (9%); *Inaperturopollenites limbatus* Balme, 1957 (6%), *Aplanopsis dampieri* (Balme, 1957) Döring, 1961 (4%), ?*Cycadopites fragilis* Singh, 1964 (3%), *Eucommiidites* Tschudy, 1962 (3%), among others). After Strivastava (1975), *Classopolis* can occupy coastal plains and well-drained mountain slopes. Vakhrameev (1981) proposed a temperate climate when percentage of *Classopolis* is between 1 and 10 %.

Angiosperm pollen are present but very rare and their determination need more investigation.

Conclusion

Palynological association in Fanar outcrop indicates a forest in proximity of the sea, in a calm estuarial or margino-littoral deposition site that was receiving regularly continental supply rich in plant debris including spores and pollen, via hydrological network. Moreover, a probable repeated rise of sea water (due possibly to tidal effect) flooding the depositional environment and bringing a considerable amount of dinoflagellates including *Oligosphaeridium* and *Spiniferites* which mingled in the deposition site.

The range of age of the different palynomorphs corroborates with admitted geological data and indicates a probable Neocomian age.

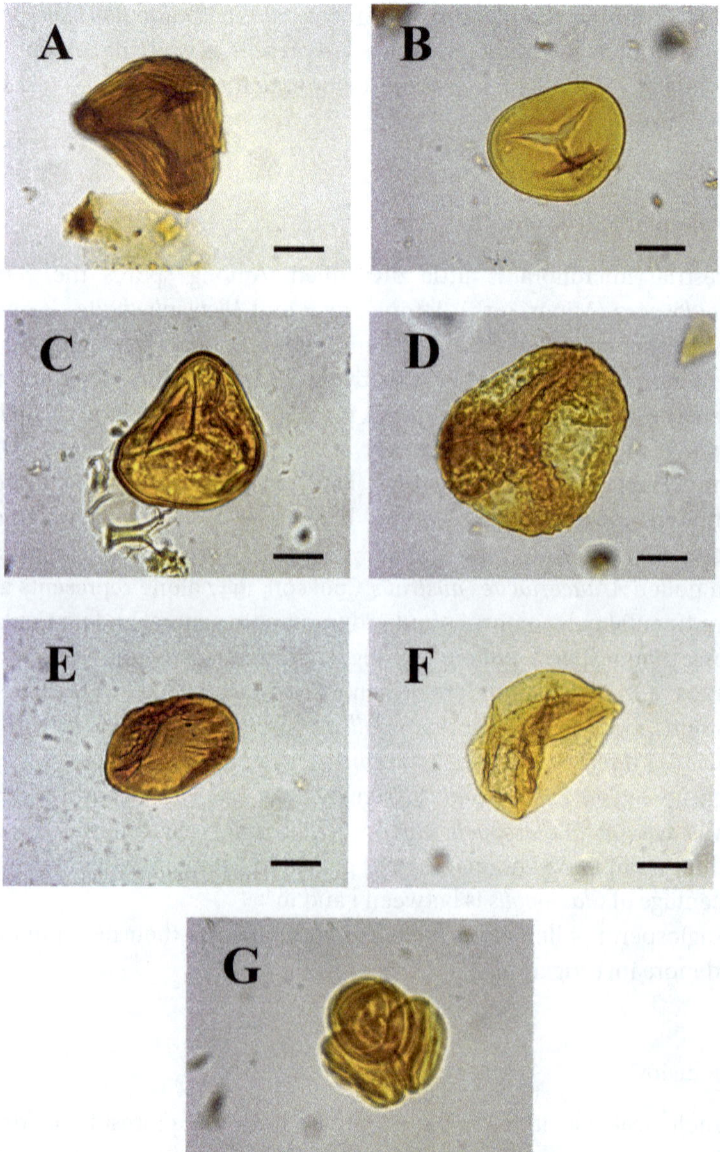

Figure 9. Terrestrial palynomorphs. (A) *Cicatricosisporites potomacensis* Brenner, 1963 (slide 8, S46 2); (B) *Deltoidospora australis* (Couper, 1953) Pocock, 1970 (slide 7, V29 0); (C) *Deltoidospora sp.* (slide 7, L51 4); (D) *Verrucosisporites sp.* (slide 8, N44 4); E) *Aplanopsis dampieri* (Balme, 1957) Döring, 1961 (slide 4, N46 0); (F) *Araucariacites australis* Cookson, 1947 (slide 4, R39 0); (G) tetrad of *Classopolis* (Pflug, 1953) Pocock et Jansonius, 1961 (slide 4, F41 3). Material is deposited in the Natural History Museum of the Lebanese University.

Figure 10. Present state of the outcrop.

Unfortunately this outcrop was very ephemeral due to the construction of buildings (Fig. 10) and we did not have enough time to rescue some material for chemical and palynological analyses.

Lebanon is a country very rich in amber outcrops that are found in Late Jurassic and Lower Cretaceous stages, among which a significant number can yield to biological inclusions. These outcrops are not officially protected and a number of them are continuously lost due to "civilization" impact. Our efforts are concentrated to classify at least some of them on the list of human heritage for their very significant scientific importance resulted from their critically noteworthy Lower Cretaceous age, origin of our Recent biota.

Acknowledgements

We would like to thank Dr Jean Dejax from the "Muséum National d'Histoire Naturelle" Paris, for his extremely helpful comments on an elder version of this manuscript. This is a contribution to the project team Biodiversity: Origin, Structure, Evolution, and Geology allotted to D.A. by the Lebanese University. Y.N. would like to thank Azm & Saadé association for partially financing his PhD.

References

Azar, D., El-Halabi, W., Nohra, Y., Maksoud, S., El-Samrani, A. & Gèze, R. (2012) New outcrop of amber from Lebanon and Syria. In: *7th Session of the World Amber Council, 18–19 May 2012*, City Council, Gdansk, pp. 18–23.

Batten, D.J. (1996) Palynofacies and paleoenvironmental interpretation. In: Jansonius, J. & McGregor D.C. (Eds), *Palynology: principles and applications, Volume 3*. American Association of Stratigraphic Palynologists Foundation, College Station, TX, pp. 1011–1064.

Dettmann, M.E. (1963) Upper Mesozoic microfloras from South-Eastern Australia. *Proceedings of the Royal Society of Victoria, N.S.* 1(77): 1–148.

Downie, C., Hussain, M.A. & Williams, G.L. (1971) Dinoflagellate cysts and acritarch associations in the Paleogene of southeast England. *Geoscience and Man* 3: 29–35.

Dubertret, L. (1945) *Carte géologique de Beyrouth* (50.000ème). Ministère des Travaux Publics, Beirut.

Eshet, Y., Moshkovitz, S., Habib, D., Benjamini, C. & Magaritz, M. (1992) Calcareous nannofossil and dinoflagellate stratigraphy across the Cretaceous-Tertiary boundary at Hor Harar, Israel. *Marine Geology* 18: 199–228.

Masure, E. (1984) L'indice de diversité et les dominances des "communautés" kystes de dinoflagellés; marqueurs bathymétriques; forage 398 D, croisière 47 B. *Bulletin de la Société Géologique de France* 26: 93–111.

Srivastava, S.K. (1975) Maastrichtian microspore assemblages from the interbasaltic lignites of Mull, Scotland. *Palaeontographica, Abteilung B* **150**: 125–156.

Vakhrameev, V.A. (1981) Pollen *Classopolis*: indicator of Jurassic and Cretaceous climates. *The Palaeobotanist* **28–29**: 301–307.

PART SIX

NEW TECHNIQUES FOR AMBER PREPARATION

A new technique for the preparation of small-sized amber samples with application to mites

Ekaterina A. Sidorchuk*

Arthropoda Laboratory, Paleontological Institute, Russian Academy of Sciences, Moscow, Russia
*E-mail: e.a.sidorchuk@gmail.com

Abstract

Small preparations are necessary to receive high-resolution morphological data on minute amber inclusions, like mites, tiny insects, pollen, fungi, etc. For mites, observations from four to six sides are often necessary for an accurate identification and systematic description. The main difficulty of such preparation is that human hand is not precise enough for holding and manipulating minute objects. Miniaturization of tools and use of holders of different kinds is necessary. This paper describes tools and protocol for routine preparation of voluminous (observable from more than two sides) amber samples of sub-millimeter size, including artificial resin embedding after vacuum treatment, trimming, grinding, and preparation for light microscopy under immersion oil. A review of received results in paleontology of amber mite inclusions is provided along with a discussion on the conservation problems raised by small size of pieces. Storage in water with thymol (preservative) is suggested, although long-time observation is yet needed to be conclusive.

Keywords

Methods; tools; optical microscopy; fossil resin; resolution; detail

Introduction

Although fossil resins and amber in particular have a long history of investigation, the development of preparation techniques and investigation methods continues (Azar 1997; Perrichot et al. 2004; Santiago-Blay 2008; Heethoff et al. 2009; Sidorchuk 2011). For medium- to large-sized inclusions, sophisticated observational methods like X-ray microtomography and nanotomography appear very promising and sometimes irreplaceable, e.g., with historical specimens (Dunlop et al. 2011a) or with opaque amber material (Lak et al. 2009; Soriano et al. 2010). For inclusions of a mite-size (ca. 1 mm), optical microscopy remains the only reliable tool to obtain morphological data, as tomography still does not give necessary detail (Heethoff et al. 2009; Dunlop et al. 2011b). The achievable resolution is highly dependent on the thickness of the resin interposed between inclusion and a microscope's lens. After my observations on Eocene European amber inclusions, the amber thickness must not exceed hundred times the size of a given feature for an accurate observation. For instance, the exact shape and position of a 10-µm-long seta will be accessible when the surface is at most 100 µm from the inclusion. For an appropriate description, a mite must be observed from at least three directions, better from six. Thus, routine work is to prepare cubes or the other six-sided pieces of sub-millimeter size. There are two problems of making such minute preparations: (1) to make flat surfaces intersecting at necessary angles, (2) to insure that inclusion itself is not damaged during the preparation. This paper describes a protocol and tools to overcome these problems, with a short discussion of means taken for subsequent conservation of such small samples.

Material and Methods

Epoxy resin embedding (Fig. 1)

For cases when enforcement of a piece is necessary (as with brittle, damaged, or unique specimens), the epoxy resin embedding procedure as described by Nascimbene and Silverstein (2000) is modified for minute specimens. OpenScience vacuum chamber designed for this purpose is a couple of plastic plates which are screw-fixed together, pot and rubber ring through which a syringe's needle is poked in (Fig. 1A). The bottom piece has a pit to accommodate the pot in which epoxy resin is mixed (I use Buehler's Epo-Thin because of its low viscosity). When mixture is made after manufacturer's protocol, an amber piece is put into the pot. The pot is covered with a rubber ring, the two halves of the chamber connected together, and the screws tightened (Fig. 1B). To remove the remaining air, the syringe is tightly attached to the tip, and the syringe's handle is pulled and fixed. For a 2-cm^3 chamber, a 20-ml syringe is normally used but a 10-ml syringe is preferred when the piece is very brittle and fissures large, so as to avoid the risk of breaking it up. The resin starts releasing air bubbles immediately, an air escapes from amber fissures and the resin replaces it (Fig. 1C), especially after the syringe is removed and pressure returned to normal. If fissures are very thin, the resin begins curing before all of them are filled. In this case one can change resin and repeat vacuum treatment several times. When the desired result is achieved, a fresh portion of resin is poured into the mold (Fig. 1D), with a size closest to that of the piece, and the piece is put inside and oriented for convenience of subsequent trimming (if needed). When the resin is cured (3 days for EpoThin), trimming and polishing can be done as described for amber.

Molds are made by pouring commercially available sanitary silicone to the mold form, which has complementary parts to shape inner and outer walls of future molds (Fig. 1E). Form parts are screw-fixed together. The bar of molds is easily detached from the form after silicone cure and cut with scissors to provide separate units (Fig. 1F).

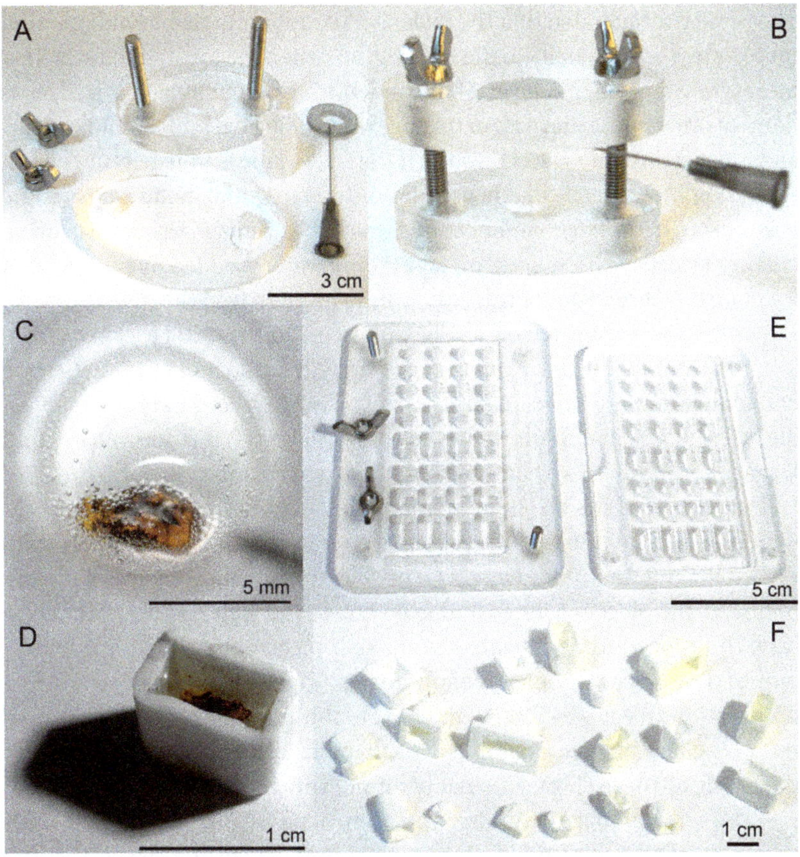

Figure 1. Vacuum treatment and artificial resin embedding of small sample. (A) OpenScience vacuum chamber, disassembled. (B) Same, assembled. (C) Epoxy resin with amber piece inside vacuum chamber, under dissection microscope while air is being released. (D) Mold with freshly embedded amber piece. (E) Form for silicone molds. (F) Silicone molds.

Trimming (Fig. 2)

Two kinds of sawing devices were tried: the use of a fixed water-fed circular saw is described in my first paper on this subject (Sidorchuk 2011), but a more efficient way is the use of a hand-held rotary tool while the sample is fixed under the dissecting microscope. If the necessary cut is larger than 1 cm^2, a Dremel 300 rotary tool with flexible drive and diamond disc is used; for smaller cuts, a Proxxon GG 12 tool is more appropriate, as the use of a stronger device for minute samples can damage them. In exceptional cases (to take away one inclusion of many), a milling cutter is useful.

To fix the sample, I use a plastic (PVC) rod with double-sided sticky tape. The piece is attached so that the inclusion to prepare is visible and oriented more or less orthogonally to the rod (Fig. 2A), then the rod and the attached piece are wrapped in a layer of one-sided transparent sticky tape. This is done to prevent the piece from turning while being cut and to hold cut-offs and most of dust and water on spot. If the sticky tape is transparent enough (I use commercially available 3M photo-safe tape), the inclusion is visible even through imperfect amber surface, as if it was moist. Using this, future cuts are traced with a marker on tape. Some water is added by syringe with a soft beak (Fig. 2B) under the tape around piece to feed the saw. Figure 2C shows the piece ready to cut. The flexible beak for the syringe is made from a standard sampler tip by pulling it over a gas flame and cutting its wide end to fit it to the syringe. In all cases described here, ordinary syringe needle is functional, but a soft tip has two advantages: (1) it is allowed as hand luggage in aircrafts; (2) it has no chance to damage the amber.

The first two cuts are done perpendicular to the rod. The rod is held with one hand, while the rotary tool is operated with the other (Fig. 2D). Before making the next two cuts, cut-offs of resin and of sticky tape are removed, the piece is rotated and fixed anew, and water is added. If necessary, the procedure is repeated to make a piece cubical or trapezoid as required by necessary observation angles (Fig. 2E) (in this case, during preparation an egg was discovered next to the female mite, and the preparation is going to have a complicated shape in order to show both). Then, the resulting piece (Fig. 2F) is ready for polishing.

If the original piece is not flat and there is no way to cut or grind it flat and parallel to one of the desired planes prior to cutting (i.e., if it is too large or contains important syninclusions), I use a soft plastic (polyurethane foam) rod with a cavity cut to accommodate the piece, the rest of the procedure runs as described, with double-sided sticky tape on the cavity bottom. Softer substances, like plasticine, paraffin and mounting wax, must be avoided because they do not prevent the piece from vibration

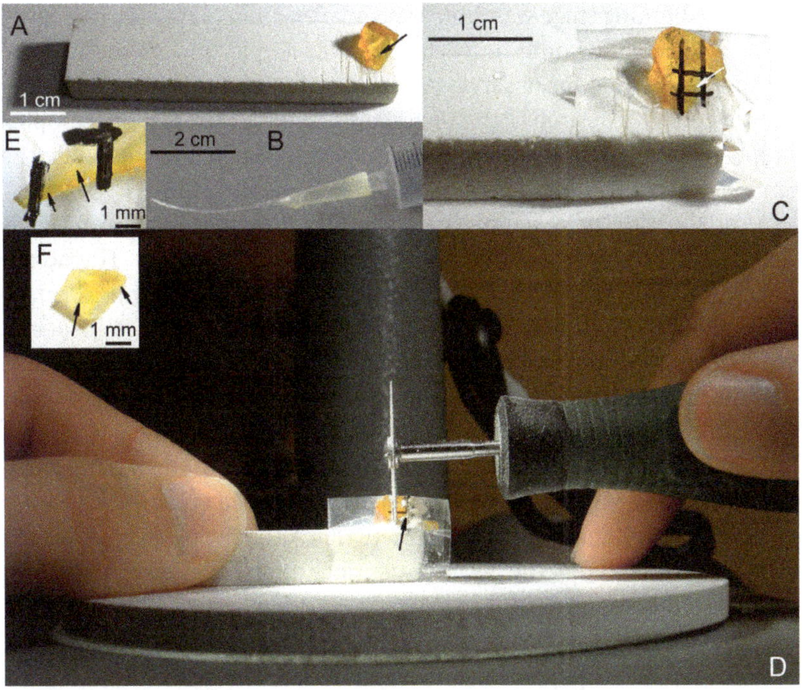

Figure 2. Trimming of small sample. (A) Sample fixed on PVC rod with double-sided sticky tape. Arrow points to inclusion. (B) Flexible syringe tip. (C) Amber piece ready for cutting. (D) Cutting the piece under dissecting microscope, arrow points to inclusion. (E) Piece ready for final cutting. (F) Freshly cut amber piece. In panels E and F, the short arrow points to an egg and the long arrow to mite inclusion.

while it is cut, so that it can tilt and be broken by the saw. Paraffin, in addition, can contaminate fissures in the piece with oily compounds while being warm.

Polishing (Fig. 3)

I currently use two polishing machines: OpenScience PollyOne (Fig. 3A) for laboratory use and a portable USB-powered MiniPolly when voyaging (Fig. 3B). Both devices use one kind of changeable and reusable discs on which sandpapers are glued, with the chamois fixed by a ring on the last disc. I normally use three (P1200, P2000 and chamois with diatomite added), sometimes P600 is necessary for larger samples. The polishing machine is water-fed with the help of a drop counter attached to a bottle with a hook to fix it over the workspace (Fig. 3C). One bottle (250 ml) is sufficient for a full day's work.

If the trimming stage was done successfully, amber layer to grind away is not more than 0.5 mm on each side of a sample, and all surfaces are oriented as needed for observation. In most cases, a cube or rectangular parallelepiped is desired as the final shape, sometimes it is more complex (trapezoidal or parallelepiped) but in any case the first task is to fix the sample flat and grind it to make two perfectly parallel sides.

After the side of a sample providing the best observation angle is chosen, the sample is fixed by this side to a holder (Fig. 3D). The holder consists of a grip and the set of interchangeable heads, each with a rectangular pit of known depth (I use 12 heads with tips from 3.0 to 0.2 mm deep). The head is chosen so that only the part of a sample to be removed remains outside the pit, then the sample is fixed by a double-sided sticky tape to the bottom of a pit in its corner to avoid any tilt. The choice is verified by applying a straight piece of polyethylene or paper to the outer surface of the head (Fig. 3E). When the piece is fixed, the head is put on the holder and grinding is done as shown in Fig. 3F. Thus, the inclusion is protected during polishing, and the resulting surface remains perfectly flat and parallel to the opposite surface (Fig. 3G, H). In between two subsequent sandpapers and after polish, the piece is cleaned in water with a soft brush or a piece of paper tissue. To detach the piece after one surface is finished, I gently push it with the fingernail from a side until it slides from the sticky tape. Most of manipulations with polished pieces are done with a rubber-tipped tweezers, described in detail by me earlier (Sidorchuk 2011). The same operation is repeated with a shallower pit to polish the opposite side.

If the final shape of sample is rectangular parallelepiped, the rest of grinding and polishing is done with the same holder, side after side.

If one or more of desired surfaces are not perpendicular to the main plane, plastic slit forceps with screw fixation (Sidorchuk 2011) are used when the other sides of the piece are already processed with rectangular fixation. The part of sample to be removed is blackened with a marker, then the piece is fixed in forceps so that only blackened part is outside, with a few micrometers of unblackened area showing up to control position of a piece under dissection microscope (Fig. 3I–K). Screw is tightened gently until the piece is fixed (not displaced by light pressure of fingernail). Grinding is done with frequent checking of the position of the piece: excessive pressure can damage the piece but unwanted rotation is possible, especially in the beginning, when the desired surface is not parallel to the existing one.

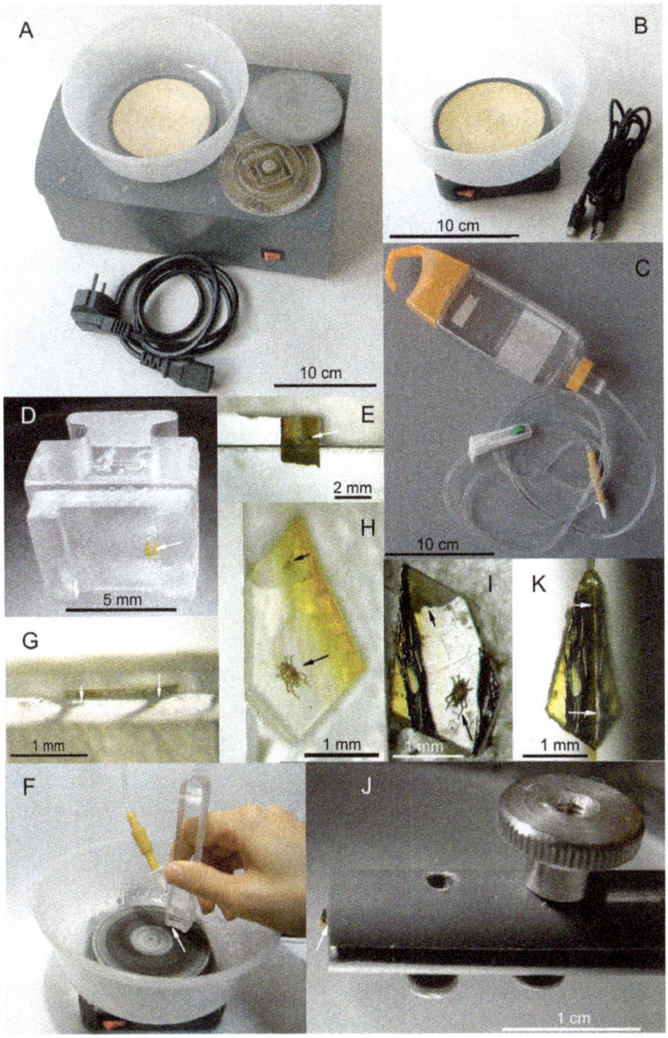

Figure 3. Grinding and polishing of small sample. (A) Laboratory-used polishing machine, OpenScience PollyOne. (B) Portable polishing machine, OpenScience MiniPolly. (C) Drop counter attached to shower gel bottle with a hook (to hang on shelves, etc.). (D) Head of OpenScience holder for flat piece polishing, arrow points to fixed piece. (E) Control of inclusion's position (polyethylene plate is below, arrow points to inclusion). (F) Position of holder during grinding and polishing, arrow points to the amber piece. (G) Control of grinding progress: piece attached to the holder's head in lateral view. (H) Piece polished from one side, top view. (I) Piece with parts to grind away blackened. (J) Same as I, fixed in slit forceps, arrow points to the piece. (K) Same as I and J, enlarged. In the panels G–K, the short arrow points to an egg and the long arrow to mite inclusion.

For final cleaning I use paper tissues soaked in water, sometimes with a drop of glycerol added to help removing residue of sticky tape and marker. Glycerol appears to be safe for Baltic, Rovno (Ukraine), Dominican Republic, Taimyr and Burmese resins. It partly dissolves India amber (surface becomes tacky), so it must be applied with caution to "new" kinds of resin.

Syringe with a paraffin plug, described in my previous paper (Sidorchuk 2011), is now extremely rarely used—only in the case of a very small and irregular piece that cannot be cut from any side, so that I have to fix it in paraffin and grind away the first surface to proceed as described above. The reasons for this abandonment are: (1) a small rotary tool when operated under a dissection microscope with a sample fixed on a rod gives higher precision than a paraffin holder (which always can be grounded away) may provide; (2) paraffin contaminates polishing surfaces; (3) working with paraffin requires hot water and ice or a refrigerator at hand (very inconvenient when working in museum collections).

Observation (Fig. 4)

The main purpose of the preparation described above is the study under a light microscope, though any other method can be applied. I mostly use cavity-slide with a drop of immersion oil or glycerol, if the amber has fissures. The cavity is half-covered with a coverslip and observation done as described by Grandjean (1949) for modern arthropods. When an immersion oil lens is necessary, a coverslip is inapplicable. In this case, I cut a stripe of polyethylene with a hole slightly larger than the inclusion and fix it with sticky tape on the coverslip attached in turn to an object slide (Fig. 4A, B). This is done to avoid excessive pressure of oil lens on the piece: albeit the lens may be pressure-protected, small amber piece can be damaged if squeezed between a firm object slide and the lens. A coverslip attached by sticky tape, on the contrary, is movable: it bents when pressed even slightly, thus protecting the specimen. The inclusion is visible through the hole (Fig. 4C, D), and the whole sample can be rotated as necessary. Normally, recently prepared inclusions contain air and are not transparent in transmitted light. After several weeks to a few months of water storage, however, they become translucent as the air is replaced by water (probably by diffusion). This gives much advantage for observation but often becomes an obstacle for taking images as an inclusion loses its shiny appearance. Such an appearance comes back after the piece is dried, which takes one or two days in air. Drying must not be done

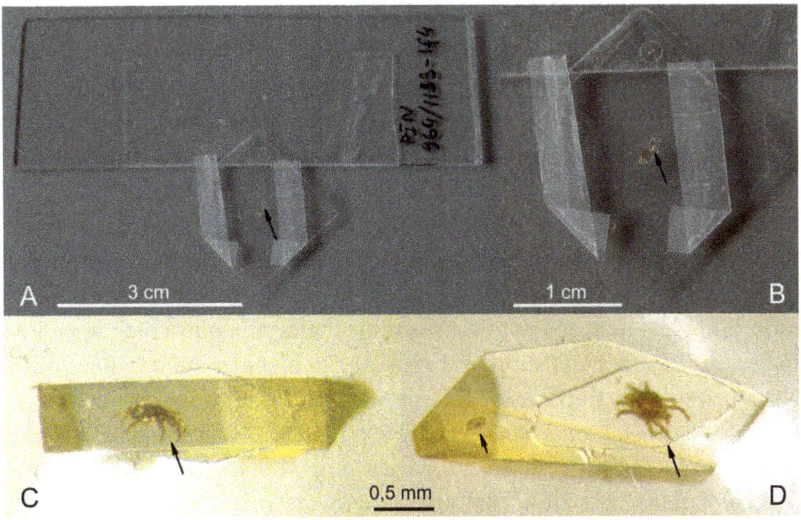

Figure 4. Mounting of amber piece for microscopy with immersion oil lens. (A) Total view of preparation, arrow points to the piece. (B) Fixation of piece on coverslip, arrow points to the piece. (C) Inclusion visible through the cut in polyethylene, arrow points to inclusion (mite in lateral view). (D) Same, dorsal view of mite. The long arrow points to the mite, the short arrow to the egg.

frequently, it is better to take the necessary images prior to water storage, as changes of relative humidity lead to deterioration of the amber surface (Bisulca et al. 2012). Illumination and some issues concerning observational conditions are discussed by Sidorchuk and Norton (2011a).

Results

The preparation of small samples allows observation of fossil mites in detail necessary for species-level identification. Light microscopy of such samples permits resolving leg and body setation, setal forms and cuticular structures of fossilized creatures less than 0.5 mm long. This permits to be conclusive about the taxonomic position of fossils and leads to exciting discoveries. For instance, we found that the mite species that is ubiquitous in Baltic amber, originally described as *Acarus rhombeus* by Koch et Berendt in 1854 and successively moved in five different families within two suborders of Acari since then, actually represents a distinctive evolutionary lineage within Astigmatina, having unique sexual dimorphism and habits, body and legs proportions and setations. These features imposed the erection of a new genus *Glaesacarus* Klimov et Sidorchuk, 2011 and a new family Glaesacaridae Klimov et Sidorchuk, 2011 to accommodate the species (Klimov & Sidorchuk 2011; Sidorchuk & Klimov 2011). Another common Baltic amber mite, *Scutoribates perornatus* Sellnick, 1918, previously thought to belong to the tropical family Eremaeozetidae, appeared to be a member of the family Unduloribatidae Piffl, 1972, now having highly disjunctive distribution in Holarctic mountains, while the generic name is a senior synonym of at least two names proposed in two different families of Oribatida. We could even resolve several morphological features not described for the family but which appeared to be common in it and support its position in higher system (Sidorchuk & Norton 2010). The third case is linking together three apparently unrelated taxons: *Plategeocranus sulcatus* (Karsch, 1884), *Strieremaeus illibatus* Sellnick, 1918, both from Baltic amber, and *Archaeorchestes minguezae* Arillo et Subías, 2000 from Àlava amber. *Archaeorchestes* appeared to be a junior synonym of *Strieremaeus*, two species being extremely similar. All three species are confamilial, thus showing that the family Archaeorchestidae Arillo et Subías, 2000 existed in Europe at least from Early Cretaceous to Eocene but seems to be now extinct (Arillo & Subías 2000; Sidorchuk & Norton 2011a, b). In all cases, resolving fine morphology permitted us to link immature and adult mites from Baltic amber, thus obtaining important data on their systematic position and some clues to their living habits.

Discussion

The small size of preparations allows fine observations but it aggravates the problem of conservation, which is very pressing for all fossil resins (Grimaldi 1993; Rasnitsyn & Quicke 2002; Bisulca et al. 2012). All observations suggest that fossil resin must be protected from direct contact with air, especially of low or unstable humidity, from high and unstable temperature, as well as from loss of volatile compounds. Therefore, airtight packaging and low temperature should be best, and embedding in medium—Canada Balsam or artificial resin or the like—has been suggested many times. Such embedding, however, leads to loss of the advantage provided by the minute voluminous sample: observation under an immersion lens of the embedded piece is not possible because of the depth, while a thin two-sided slide mount does not allow observation from the other four sides of the specimen. Moreover, Canada Balsam preparation, when sealed prior to complete evaporation of solvent, is not safe for amber piece: i.e., some pieces of Lebanese amber with mites, which were made and sealed several years ago, now contain disintegrated amber pieces with most of the inclusions irreparably damaged, with only the cuticle remnants conserved and visible (author's observations and communication by M. Judson, 2011).

Archaeological finds of well-preserved polished objects in water-saturated sediments, e.g., a Mesolitic amulet from Halleby River in the western part of Zealand Island in Denmark which retained its bright yellow colour for over at least 4000 years (Fraquet 1987), suggest water storage of amber specimens might be suitable for long-term conservation. The problem is to avoid evaporation and biological activity in water. Also, amber pieces that were found within lignitic sediments are often enriched in pyrite (FeS_2), so storage in water must be avoided for those amber pieces to prevent from oxidation of the pyrite. I currently use paraffin or Parafilm-sealed Eppendorf tubes with an aqueous solution of thymol (preservative), stored at +7°C, but time is too short to be conclusive about results (no change of colour or surface, no biological activity observed in Baltic amber specimens over three years). Unique, flat and damaged specimens are mounted in EpoThin between two coverslips, some were slide-mounted in Euparal and sealed with Red Insulating Varnish.

Acknowledgements

I am very grateful to the colleagues who provided discussion and advice at different stages of this work: A.P. Vlaskin (Schmalhausen Institute of Zoology, National Academy of Sciences of Ukraine, Kiev, Ukraine), Prof. A.P. Rasnitsyn, Drs. I.D. Sukacheva, D.S. Kopylov and D.V. Vasilenko, A.G. Vlasov (Borissyak Paleontological Institute RAS, Moscow, Russia), Dr. D.D. Vorontsov (Kol'tsov Institute of Biology and Development, RAS, Moscow), Prof. B. Kosmowska-Ceranowicz and Mgr. K. Kwiatkowska (Museum of the Earth, PAN, Warsaw, Poland), Dr. E.E. Ezhova (Shirshov Institute of Oceanology, Atlantic Branch, Kaliningrad, Russia), Prof. W. Weitschat (Geologisch-Palaeontologisches Institut, Universität Hamburg), Prof. R.A. Norton (SUNY College of Environmental Science and Forestry, Syracuse, NY, USA), Dr. A. Arillo (Universidad Complutense, Madrid, Spain), Drs. A. Nel and M. Judson (Muséum National d'Histoire Naturelle, Paris), Prof. D. Grimaldi and Dr. P. Nascimbene (American Museum of Natural History, New York, NY, USA (AMNH)), Dr. Elżbieta